Josef Barla
The Techno-Apparatus of Bodily Production

Science Studies

Josef Barla is a postdoc researcher in the Biotechnology, Nature and Society research group based at Goethe University Frankfurt. He studied Sociology and Philosophy at the University of Vienna. His research focuses on questions at the intersection of technology, ecology, (techno-)biopolitics, and care.

Josef Barla
The Techno-Apparatus of Bodily Production
A New Materialist Theory of Technology and the Body

Bibliographic information published by the Deutsche Nationalbibliothek
The Deutsche Nationalbibliothek lists this publication in the Deutsche Nationalbibliografie; detailed bibliographic data are available in the Internet at http://dnb.d-nb.de

© 2019 transcript Verlag, Bielefeld

Cover layout: Maria Arndt, Bielefeld
Cover illustration: »Gamochonia. – Trichterkraken« (Detail, image edited by Ines Handler), Copyright: Giltsch Adolf and Ernst Haeckel; CC0
Printed by Majuskel Medienproduktion GmbH, Wetzlar
Print-ISBN 978-3-8376-4744-0
PDF-ISBN 978-3-8394-4744-4
https://doi.org/10.14361/9783839447444

Table of Contents

Acknowledgments | 7

Introduction | 9

1 Mapping the Terrain | 19
Technology Beyond Determinism | 19
The Technological is Political 25
Technology and the Body: An Asymmetrical Relation | 38

2 Locating the Technological with/in Rhizomatic Networks | 57
The Mode of Existence of Technical Objects | 59
Technical Mediation as Processes of Mutual Mobilization | 57
Power and Agency in Heterogeneous Networks | 71
Relational Ontology and the Question of the Political | 80
The Absent Present Body | 92

3 Re(con)figuring the Apparatus | 101
On Material-Semiotic Actors and Generative Nodes | 102
Apparatuses as Boundary-Drawing Practices | 124
Figurations Matter: The Techno-Apparatus as Figure and Method | 144

4 Cutting Technology and the Body Together-Apart | 151
Difference that Matters: The Biopolitics of the Spirometer | 152
Materializing Authentic Bodies The *Human Provenance Pilot Project* 168
Conclusions: A New Materialist Theory of Technology and the Body | 182

References | 209

Acknowledgments

As an entanglement itself, this book would not exist if it were not for the help, ideas, and contributions of friends, colleagues, and many other things of material and immaterial nature. Setting up a list cannot do justice for the friendships, intellectual and personal encounters, and the companionships that made it possible to write this book over the last few years – which went by all too quickly. Nevertheless, I wish to thank all those minds and hearts that inspired and encouraged me to write this book, hoping that one way or the other they will find themselves in it.

I would like to express all my gratitude to Mona Singer who introduced me into the world of feminist science fiction and from there to the fabulous work of Donna Haraway in the first place. More than for her patience, I would like to thank her for profoundly shaping my critical thinking and for constantly reminding me that 'we' are always already in the thick of things. Thank you also for the many meetings and the highly productive discussions we had in the doctoral seminar group.

I am deeply grateful to Karen Barad for inviting me as a visiting fellow to the Science and Justice Research Center at the University of California at Santa Cruz in the academic year of 2012/2013. Thank you for establishing our small theory group and for introducing me to Magdalena Górska, Bue Thastum, and Elaine Gan – I still miss the discussions we had at campus as well as our inspiring meetings off campus. I would also like to thank Jenny Reardon and Andrew Mathews for welcoming me warmly and for providing me with a workspace at the Science and Justice Research Center as well as my other colleagues at Oakes College for all the stirring debates we had.

My profound thanks go to the faculty of the Initiativkolleg "Gender, Violence and Agency in the Era of Globalization" (GIK) at the University of Vienna – Nikolaus Benke, Eva Flicker, Susanne Hochreiter, Elisabeth Holzleithner, Eva Kreisky, Elke Mader, Maria Mesner, Birgit Sauer, Mona Singer, and Sabine Strasser – for making my work and those of my fellow colleagues possible by initiating the GIK (a process that took several years and consumed a lot of energy and resources), as well as to Sigrid Schmitz who not only shared the office with us but also her ideas and expertise on our projects. I would also like to thank Cecilia Åsberg and Iris van der Tuin for their generous reviews and highly valuable feedback on an earlier draft of this book.

I wish to thank all my friends who supported me by reading and commenting on drafts of this manuscript, for providing me with invaluable feedback that strengthened my work, and for reminding me that there is, and should be, a world beyond academia. Thanks especially to Isabella Amir, David Hoffmann, Christoph Hubatschke, Susi Kimm, Mathis Kronschläger, Katharina Maly, Anna Petran, Louise Thiel, Ina Tiefenbacher, and Sam Wade for reading my work and for providing me with helpful insights and remarks that enriched my draft.

Finally, I am especially indebted, once again, as always, to Ines and Maggie who gave me courage to pursue the project when I lost faith in it.

Introduction

> What we need is to make a difference in material-semiotic apparatuses, to diffract the rays of technoscience so that we get more promising interference patterns on the recording films of our lives and bodies. Diffraction is an optical metaphor for the effort to make a difference in the world.
>
> —Donna Haraway/*Modest Witness*

Over the last few decades, the idea that technology is not the Other of the body has become more and more pervasive in poststructuralist and posthumanist theories. Jacques Derrida, for example, reminds us that "there is no natural, originary body: technology has not simply added itself, from the outside or after the fact, as a foreign body" (Derrida 1995: 244). For Derrida, technology is "'originally' at work in place in the supposedly ideal interiority of the 'body and the soul'" (ibid). Similarly, Donna Haraway argues that "technologies are not mediations, something in between us" but rather "what Merleau-Ponty called 'infoldings of the flesh.' What happens in the folds is what is important." (Haraway 2008: 249) Both Derrida and Haraway seem to suggest that technologies and bodies are inextricably entangled with one another. Technologies are to be understood as always embodied technologies while material bodies have to be considered as always technologized bodies. What remains unclear, however, is the question as to how to understand the embodiment of technologies, and how to make sense of the role of technologies and technoscientific practices in the processes of the calling into being of particularly materialized bodies.

Discourse theories and especially the concepts of performativity and interpellation have been crucial tools for understanding how and which subjectivities and consequently also bodies come to matter, but they also remain limited. Judith Butler, for example, provides us with a robust account of how regulatory practices performatively produce the bodies they govern by hailing them into being. Instead of referring to "a singular or deliberate 'act'", the notion of performativity, here, denotes "the reiterative and citational practices by which discourse produces effects that it names" (Butler 1993: 2). Precisely because for Butler matter is "a process of materialization that stabilizes over time to produce the effect of boundary, fixity, and surface" (ibid: 9), bodies are not given and fixed entities with clear-cut boundaries and properties but have to be understood as "the effect of power, as power's most productive effect" (ibid). Far from reducing bodies to arbitrary products of language, Butler gets to the heart of the very materiality of bodies. However, in an important sense, the material dimensions of the regulatory practices through which bodies are performatively enacted remain undertheorized.[1]

Departing from these insights, this book engages with the question of how to understand the material dimension of regulatory practices from a perspective that regards bodies as always already technologized bodies and technology as always already a part of 'us'. Stories about the power of technologies and techno-scientific practices to shape, discipline, objectify, and even determine the body have been and still continue to be of fundamental importance for critical analyses of social power relations and inequalities. But what if it is not only too easy but also too limiting, both theoretically and politically, to assume that technologies function precisely according to particular social power relations and interests that have been inscribed into them, disciplining and objectifying a mere passive body? What if, what is needed in critical approaches to technology and the body is also a story of technological failure instead of the same old tales of the domination and disciplining of the body through the means of science and technology?

Feminist theorists such as Donna Haraway, Rosi Braidotti, Marilyn Strathern, and many others have stressed the need for new stories, new methods, and new figures to emerge in order to arrive at better understandings of our past, present, and future in a world that has been thoroughly transformed by techno-biopower and biocapitalism. Rather than mirroring an otherwise mute and inaccessible world, figures and concepts function as promising tools for actively en- and revisioning more liveable futures in a common world. Figures and concepts are thus always material and embodied, they are always part of the world in its becoming. The concept of the apparatus of bodily production is perhaps one of the most powerful tools that allows for new stories about technologies and

---

1 Cf. Karen Barad's productive engagement with Butler's notion of performativity in *Meeting the Universe Halfway* (2007), as well as chapter three in this book.

bodies in their entanglement with one another to emerge, contesting a thinking that differentiates between original and (illegitimate) copy, the natural and the artificial, dynamic mind and passive ahistorical matter.

What lies at the heart of this book is the attempt at a return to the figure of the apparatus of bodily production in order to take it a step further. As the potential of figures comes from "the join between the figurative and the factual" (Haraway 2000: 24), my aim lies in outlining a new materialist notion of the apparatus of bodily production as both a figure that allows us to shift our attention to concrete sites where biological, technological, social, economic, and political forces intra-act and in doing so mutually materialize a particular phenomenon, and as a speculative tool for engaging with questions of power and becoming that pervade those sites. In doing so, a deeper understanding of how not only particular knowledges about bodies but also specifically reconfigured bodies come to matter in the context of the technosciences shall be developed. Instead of already starting with an understanding of the body as a mere object or as a surface on that powerful technologies and technoscientific practices act upon, the question for the generative and unruly potentials of bodies will accompany this book.

Employing the concept of the techno-apparatus of bodily production as a speculative tool for the taking into account of the regulatory practices through which bodies materialize along with their boundaries and meanings, this book will put forward an understanding of the relation between the body and technology as a relation of indeterminacy. As an ontological concept, the notion of indeterminacy is not about a lack of knowledge, but rather about spatio-temporal undecidability. That is, instead of designating that 'we' cannot know where the body ends and technology begins, the notion of indeterminacy highlights that it cannot be determined in advance and once and for all, where the boundaries between bodies and technologies run. It is only through particular material-discursive practices (in which not only humans but also nonhumans take part) that particular boundaries and properties manifest and become meaningful. From this follows that bodily being is always technologized being and technologies always embodied phenomena. In an important sense, however, such an account does not intend to conflate technologies and bodies, or to erase differences, as well as questions of accountability and responsibility. On the contrary, it implies that both technologies and bodies are always themselves multiplicities or assemblages. In doing so, it affirms not only that bodies and technologies are entangled with one another in multiple ways, and with very different ethical and political consequences, but also the fact that neither technologies nor material bodies can be understood ahistorically and static. Indeterminacy "does not mean that there are no facts, no histories, no bleeding – on the contrary, indeterminacies are constitutive of the very materiality of being", Karen Barad (2014: 177) reminds us. As "a mode of existence" (Latour 2002; Simondon 2016), technology is at the heart of questions of mattering. As Haraway reminds us, technologies are about particular ways of making intimate connections, and it "matters which ones get made and unmade" (Haraway in Kunzru 1997). That is, it matters whether 'our' intimate connections with technologies are aiming at the appropriation and commodification of the generative forces of human and nonhuman bodies for the demands of the political economy, or working toward the realization of more livable futures for everyone.

This book is divided into four parts. Following Deleuze and Guattari's call to produce maps rather than to draw exact copies, the aim of the opening chapter is twofold. First, I will critically engage with pivotal sociological and philosophical debates on the relationship between technology and the body. Since mapping is never an innocent practice but always situated, I neither want to produce a linear history here but rather reconstruct genealogies, nor do I want to present the produced map as the only possible one, as it is only one story and how it could be told. In doing so, I will argue that while social constructionist approaches to technology were highly successful in contesting technological determinism, material bodies in their multiple and dynamic entanglements with technologies had been largely ignored. That is to say, while the dichotomy of technology and society has been contested, the dichotomy of technology and the body, and with it the dichotomy of the natural and the artificial, the organic and the mechanical, remained largely untouched. As I shall point out in this chapter, feminist scholars were not only the ones who highlighted a number of problematic blind spots with regard to questions of power within this body of work, but also those who brought the lived, material body back in. Despite the fact that the body has been conceptualized in a number of highly insightful and productive ways, as soon as the question of technology comes in many of these approaches tend to understand the relationship of technology and the body as one in which the latter remains a mute, disciplined and disempowered object or even a mere surface on that technologies and technoscientific practices act upon. This is particularly, but by far not only, true for analyses of new reproductive and biomedical technologies. As a consequence not only the relation of technology and the body is understood as a relation of connection rather than one of entanglement, but the dichotomies of the natural and the artificial, nature and culture, interiority and exteriority, active and passive, human and nonhuman are reinforced as well. Moreover, while the body remains a silent and passive one, a body that is constantly threatened, meaning, disciplined, surveilled, fragmented, and fundamentally reworked by particular technologies and technoscientific practices, technology tends to become a placeholder for social power relations, and hence mere politics by other means. In doing so, as I will problematize, many of the approaches within this framework seem to assume that there is almost a guarantee that technologies function precisely according to certain individual or collective interests and social norms that have been inscribed into them.

Second, against this backdrop, I will argue that what gets lost in many of these understandings of the relationship between technology and the body is nothing less than the possibility to understand how bodies materialize in their entanglement with political, technological, technoscientific, biological, economic, and other forces, precisely without understanding the bodies involved as mute and passive objects. Rather than denying the effects of power and technologies on material bodies, I argue that what is needed are theories, concepts, and methods that allow us to take into account not only the facticity of material bodies but also their potentials to be unruly and to 'kick back'; or in other words, following Deleuze's reading of Spinoza the question has to be what *bodies* are *capable of* rather than what *the* body *is*. Such an approach might promise a deeper, or at least a fundamentally different, understanding of how and why technologies can also always fail in their attempts at reconfiguring bodies for the demands of the political economy. Traversing narratives which understand material bodies either as brute passive matter or first and foremost as effects of powerful discourses and technologies (including technologies of the self), such an approach has to start by acknowledging that there is no such thing as *the* body but only bodies in the plural form, which are always assemblages themselves. What is more, such an approach also demands for a reworked understanding of technology; an understanding that regards technologies as clearly political and yet without equating them with mere politics by other means, allowing for new stories not only about technologies but also material bodies and agency to emerge.

Drawing on Bruno Latour's notion of technology as a process rather than a product or a mere artifact, in chapter two, I put to the fore an understanding of *technology* denoting a particular mode of being and knowing, a mode of engaging with our world as part of it, while *technologies* in the plural form will refer to relational matrices, that is, to specific worldly constellations of mobilizations and reconfigurations which happen between heterogonous actors. I will point out in this chapter that Latour's substantially through the philosophy of Michel Serres and Gilbert Simondon influenced notion of technology as a process that "liquefies all things and at the same time gives them new durability, solidity, consistency" (Latour 2013: 225) might allow us to go well beyond narratives that circle around an understanding of technology either as mere means to an end or first and foremost as materialized and condensed social relations. Such an account not only shifts the view from objects to relations, from fixed and static arrangements to fluid becomings, but also suggests that rather than talking about technology in the singular form, inquiries always have to take into account *technologies* in their multiple entanglements with other social, culture, biological, economic, and scientific forces. However, while such a reading of Latour's philosophy can be regarded as successful in providing us with a relational and processual understanding of technologies, Latour has surprisingly little to say about material bodies. As I point out, the body only appears either in the context of biological or even organicist metaphors for "science in the making", or as an entity that is being affected. Nevertheless, it is this second idea which is highly influenced by Spinoza and Nietzsche that not only promises an understanding of the body as always being entangled with other bodies, and therefore as always an assemblage itself, but might provide us also with an understanding of bodily activity or agency which breaks with the dichotomy of active and passive, subject and object, individual and collective. It does so in so far as it regards the body as an entity that is always put into motion, "meaning 'effectuated'" or "moved", as Latour (2004a: 205) states, by other bodies. Bodily agency, therefore, is something that emerges only *in* and *through* entanglements with other bodies (human and nonhuman ones alike), rather than being something that bodies as distinct entities already possess. I conclude this chapter by arguing that not so much the relationship between technology and the body, but rather technologies and bodies as parts of particular apparatuses are to be taken into account.

Chapter three elaborates this argument, constituting the theoretical and methodological heart of this book. To this end, this chapter follows Donna Haraway's and Karen Barad's call for developing diffraction apparatus in order to be able to study bodies and technologies in their manifold entanglements with one another. In doing so, I first engage with Donna Haraway's figure of the material-semiotic actor as a generative axis of the apparatus of bodily production, defending it against a number of false accusations and misreadings. At the same time, I argue that even though Haraway has much to say about the material-semiotic actor, the figure of the apparatus of bodily production itself remains rather underdeveloped, and thus conceptually and methodologically vague. In what follows, I demonstrate that by drawing on Haraway's insights the feminist quantum physicist Karen Barad further elaborates the concept of the apparatus of bodily production in an important sense. Reading quantum field theory (in particular, the works of the quantum physicist Niels Bohr), feminist epistemologies, and poststructuralist theories diffractively through one another, Barad develops a posthumanist account of performativity that not only allows for a deeper understanding of the epistemological but also of the ontological dimensions of technoscientific and other practices. Discussing Barad's key concepts, I point out in this chapter that her account of performativity and materialization not only brings to the fore the ethical dimensions of becoming, but might also provide us with the foundations for an actualized understanding of technobiopolitics. It is precisely Barad's reworked understanding of apparatuses as generative material-discursive practices (in the sense of material, embodied concepts rather than as instruments which are limited to laboratories or other places), as I will demonstrate, that promises to be of fundamental value for a reworking of the concept of the apparatus of bodily production.

Instead of mirroring reality or functioning as mere metaphors, concepts work toward "the dramatization of processes of becoming" (Braidotti 2013: 164). Hence, in what follows, my aim in this chapter lies in outlining a concept of the techno-apparatus of bodily production as both a figure and a speculative tool. While as a figure, the techno-apparatus of bodily production refers to sites where human and nonhuman, biological, technological, social, political, and economic forces intra-act and in doing so performatively enact specific phenomena whose boundaries and properties cannot be separated from the very apparatus of bodily production through which they came to matter, as a speculative tool, it functions as a mapping practice for the analysis of material-discursive practices through which bodies along with their boundaries and meanings come to matter. These two aspects of the concept the techno-apparatus of bodily production cannot be separated from one another. Neither can the fact that as a situated researcher one is always part of the world one seeks to understand through apparatuses of knowledge production; rendering one accountable for the boundaries, the inclusions, and the exclusions at stake.

The concluding chapter, chapter four, ties together the insights and arguments developed in the book by turning to two concrete "worldly examples", as Donna Haraway (2000: 46) would put it, exploring them as techno-apparatuses of bodily production. Drawing on Lundy Brown's groundbreaking work "Breathing Race Into the Machine. The Surprising Career of the Spirometer from Plantation to Genetics", in the first part of this chapter, I turn to the spirometer, a medical device for measuring lung capacity and 'vitality' with a deeply racist history. In contrast to Brown however, I argue that, rather than representing a mere technical instrument in which race and social inequality somehow have been inscribed into, it might make more sense to understand the spirometer as a techno-apparatus of bodily production – that is, as an assemblage consisting of technical, scientific, medical, economic, political, and biological entities, relations, and forces. In doing so, I demonstrate that understood as such a techno-apparatus of bodily production, the spirometer can be taken into account not only as producing the very phenomenon it seeks to measure, namely, 'vital capacity', but also as performatively enacting the corresponding bodies, marking black bodies, due to their alleged lower vitality, as less effective and resilient 'by nature' compared to their white counterparts. Thus, in what follows, I put to the fore an understanding of vital capacity as the material-discursive effect of the bodies, technologies, politics, discourses, social relations, and the environmental influences involved; rather than being only a brute natural fact or a mere social construction. I conclude that such an account demonstrates that measurements have not only epistemological but also ontological, and therefore ethico-political consequences in determining who is eligible for compensation payment when it comes to disability and serious illness and who not, whose lives matter and whose not.

The second part of this chapter focuses on the *Human Provenance Pilot Project*, a project initiated by the UK Border Agency with aim to combat undocumented migration with the means of new biometric technologies and technoscientific practices, targeting primarily the bodies of African asylum seekers. The project operated following the assumption that bodies cannot lie, and neither can isotopes and genes. Consequently, the UK Border Agency believed that a combination of DNA ancestry testing along with strontium isotope analysis could reveal an applicant's "true country of origin" (UK Border Agency 2009a). Analyzing the project as a techno-apparatus of bodily production, I demonstrate that despite its aim, the project has not so much revealed supposed truths about the bodies tested but rather sought to technoscientifically enact what counts as an authentic body that is bound up with particular geographies, geologies, and political images. What is more, the bodies the project sought to materialize were not just any bodies but bodies marked by ethnicity, nationality, and race, illustrating, as I argue, that it would be a mistake to believe that with/in genomics race would become less meaningful on a subdermal or molecular scale.

In both cases, as I display, the concept of the techno-apparatus of bodily production as a figure and a speculative tool not only provides us with an understanding how apparatuses function as material-discursive regulatory practices, enacting particularly re(con)figured bodies, but also suggests how these apparatuses failed in doing so, in so far as in neither of the cases discussed the bodies and technologies involved behaved as expected. While in the case of the *Human Provenance Pilot Project* the bodies tested could not be addressed as mere passive objects of knowledge or even as information storage devices that could be read like an 'open book', difference in vital capacity in the context of the use of the spirometer becomes evident as neither a natural biological fact, nor as a social construction, but rather as the material-discursive effect of technological, biological, political, cultural, and economic forces. It is precisely for this reason that naturalizing the variability in vital capacity in living, breathing bodies in order to achieve a certain political affect, proved and continuous to prove to be difficult.

Even though technology as a mode of being is inseparable from 'us', the convergence of novel bio-genetic and information and communication technologies not only produces more intimate entanglements between technologies and bodies, the artificial and the natural the digital and the analog, but also calls for novel figures and concepts for analyzing the ethico-political consequences of these very entanglements with regard to what counts as matter and how matter matters. Instead of departing from the idea that technologies are either not much more than mere tools that can be used for liberating or oppressing purposes, or an exterior/ized Other to material bodies, constantly threatening them in their integrity and supposed 'naturalness', this book puts forward the idea that what a particular technology or a particular body can do remains invisible or undetermined as long as 'we' understand both as separate, distinct entities. Only if 'we' shift our attention to the techno-apparatuses of bodily production as performative practices of simultaneously generating matter and meaning, we might be able to learn what concrete bodies and technologies in their entanglements with one another can do. Therefore, rather than trying to get rid of technology, 'we' need to integrate it even more deeply in our concepts of subjectivity. Contributing to ongoing debates in new materialist and posthumanist theories of technology and the body, this book aims at a deeper understanding of the processes of the materialization of particularly re(con)figured bodies in the context of the techno-science, and thus at a more materialist understanding of performativity.

1 Mapping the Terrain

> If talk of the human body falls prey to mechanical materialism, then one will find oneself in need of some spiritualistic or psychologistic language in order to cope with everything that such a view cannot account for. It is in this sense that vulgar materialism breeds idealism. The more, in such an epoch, the body is reduced to one object among others, the more overweening will wax the subjectivity which tries to compensate for this humiliation.
>
> —Terry Eagleton/Self-Undoing Subjects

TECHNOLOGY BEYOND DETERMINISM

Technological determinism had been a powerful narrative informing our understanding of technology for decades. Notions such as the Industrial Revolution or the Digital Revolution imply that technology not only plays a crucial role in social change but also that new technologies may even cause new historical epochs. The idea that technology is an autonomous force outside social control and yet at the same time also the prime agent of social change led to the belief that technology would not only determine society but also our very lives and bodies almost every moment and everywhere. By the second half of the 20th century, however, the power of this narrative has been broken. How could it be possible that artifacts produced by humans in their very actions have their own intrinsic logic, independent of social influences? The idea of technology as the subject of history, as well as the belief that technological change would inevitably lead to social change, soon seemed to be too simple.

The effects of technologies on social life are much more complex and less certain than most technological determinist theories described. However, despite the fact that technological determinist approaches provided an oversimplified theory of the relationship of technology and society, it is hard to deny that technologies can have far-reaching effects on our lives and the ways we live together. It is hard to deny that technology matters. It was precisely this assumption that led the way to a number of social constructionist[1] approaches to technology, which emerged in the wake of the Strong Programme of the sociology of scientific knowledge (SSK) in the late 1970s. Even though Ludwik Fleck (1935) had already highlighted the process of knowledge production as a social activity, it was Ludwig Wittgenstein (1953) who shifted the focus within the philosophy of science from the question of 'how the world actually *is*' to the question of 'how the world is *represented* in language and knowledge'. Influenced by Wittgenstein's philosophy early works in the sociology of scientific knowledge (see, for example, Barnes 1974; Bloor 1976) sought to empirically reframe the, until then mainly philosophically examined, question of the nature of scientific knowledge.

Scientific realism not only assumes the existence of a world 'out there' independent of its observation (and observers), but also the possibility of establishing a correspondence between the world as it is and our perception of it. Knowledge, hence, mirrors reality. Following this logic, successful theories were successful precisely because they successfully mirrored reality, whereas unsuccessful theories failed because they were simply false, that is, flawed accounts of reality. Through studying the work of scientists in action, the SSK explored how the concepts and theories of scientists are constructed and how, in turn, those concepts and theories structure the reality the scientists seek to understand. Field

---

1   The terms constructionism and constructivism are often used interchangeably. While *constructivism*, however, refers to learning theory and theories of knowledge in the tradition of Jean Piaget's work on cognitive development in particular, the term *constructionism* can be traced back to Peter Berger and Thomas Luckman's *The Social Construction of Reality* (1967), in which they argue that the world as well as our knowledge about it is not found somewhere 'out there' but rather actively constructed through human practices and interactions, that is, *socially*. Rooted in Kantian philosophy, social constructionism eschews determinism, naturalism, and fatalism by highlighting that no such things as pure natural facts would exist. Although not building a homogenous group, social constructionist theories can be defined as approaches that "aim at displaying or analyzing actual, historically situated, social interactions or causal routes that led to, or were involved in, the coming into being or establishing of some present entity or fact" (Hacking 1999: 48). When I use the term social constructionism in the following pages, I refer to approaches in the tradition of this thinking.

research in laboratories and at other places of scientific knowledge production brought the SSK to the conclusion that scientific knowledge production has to be understood as a social practice, as social action. Scientific knowledge, thus, is socially fabricated precisely because science itself is a social practice. Economic and political constraints, funding considerations, social relations, moral and religious beliefs, bias, and much more, not only shape the ways knowledge is produced but also the actual outcome of scientific inquiries. Therefore, scientific knowledge would not be fundamentally different from other forms of knowledge or more true than other forms of knowledge.[2] It is 'only' for historical and cultural reasons, because the notion of knowledge has been tied closely to the notion of truth for centuries, that science is associated with truth. This being so, truth becomes relativized to a historically, culturally, and spatio-temporally situated social construction. Knowledge and truth are what at a particular time and at a particular place is accepted as precisely that. Since all knowledge has to be treated as being socially constructed, there is nothing epistemologically singular about scientific knowledge. Nevertheless, it would be a misunderstanding of the Strong Programme to think that this would mean that "all believes are equally true or equally false" (Bloor 1991: 166). Rather, "all beliefs are on a par with one another with respect to the causes of their credibility" (Barnes/Bloor 1982: 23). Thus, the fact that knowledge is socially constructed does not mean that "knowledge depends *exclusively* on social variables such as interests" (Bloor 1991: 166). What followed from this conclusion was not only that science as human practice have to be subject to sociological analyses but also that, precisely because science is a human practice, sociologists who seek to understand the nature of scientific knowledge have to avoid explaining truth as a direct input from the world 'out there', as something that can be explained only in realistic terms, whereas failures are ascribed to societal influences and therefore have to be explained sociologically. In lieu of this reductionist explanation, Bloor stresses that "truth and falsity, rationality and irrationality, success and failure [...] require explanation" (Bloor 1976: 5) using the same type of cause.

---

2   Nevertheless it is important to note here that Michel Foucault reminds us to bear in mind that as a regime of truth science is much harder to stabilize than other discourses, requiring a gigantic effort. A vast number of technologies, practices, and actors are involved in the stabilization of scientific knowledge. Furthermore, understanding science as a human practice alone also ignores that nonhuman entities (including instruments, apparatuses, and technologies) may not only play an active but also a crucial role within processes of scientific knowledge production. I shall return to this idea in the following chapters.

However, the focus here lies exclusively on epistemological questions leaving ontological ones aside. As I shall elaborate later in this chapter this leads to the problem that while the SSK and many social constructionist approaches building on this body of work claim that even though there is a reality 'out there', there is no way to get through to this reality. Our perception of reality is always and necessarily mediated, hence there are no such things as natural facts. And yet, in an unsettling way, there still remains a reality 'out there' hidden behind or beyond its socio-cultural and linguistic construction which leads to a number of epistemological and ontological difficulties.

Despite the fact that the sociology of scientific knowledge examined the role of apparatuses, machines, and measuring instruments in the process of knowledge production (see Shapin/Schaffer 1985; Collins 1995; as well as in parts Collins/Kusch 1998) this body of work had little to say about technology itself. It was not until the programmatic article *The Social Construction of Facts and Artifacts* by Trevor Pinch and Wiebe E. Bijker (1984) that technical objects came into the focus of social constructionist theories. Pinch and Bijker took up the assumptions and insights of the Strong Programme and reframed them for a sociological analysis of technology. In doing so, they not only contested the until the 1970s dominant idea that while science is about the discovery of truth, technology is about the application of truth, but also argued that technology does not "evolve under the impetus of some necessary inner technological or scientific logic" (Bijker/Law 1992: 3). Scholars within this body of work shifted the focus toward the question how innovation and technological development can be understood. With this shift the relationship between technology and society moved to the center of attention. Soon the linear model of technological innovation that implied that scientific research leads directly to the development of new technologies which again lead straight to the production and ultimately to the usage of technical objects had been replaced with a multidirectional model within which social groups are not mere users of technical objects but an obligatory passage point in the phase of the design, production, and acceptance. Technical objects gain their meaning through the complex trade-offs between different social groups and institutions following different interests. By arguing that technological artifacts emerge "out of conflict, difference or resistance" (Bijker/Law 1992: 9) carried out by social groups with different interests, this approach to technology broke with the idea that technologies are the result of sole (usually male) inventors. The history of science and technology not only shows that the invention of technologies was often the achievement of groups, collectives, and institutions, rather than of sole inventors, but also that the narrative of the male genius usually renders invisible the contributions of female scientists and engineers. Catherine Green, for example, invented the Cotton Gin, a machine to separate cotton fibers from their seeds in the late 18th century. Emily Davenport helped to develop the first fully functioning electronic motor together with her husband. Stephanie Kwolek developed a synthetic fiber that became known under the name Kevlar in the late 1960s. The same is also true for scientific discoveries. Rosalind Franklin's involvement in the discovery of the double helix structure of DNA in the early 1950s is a famous example in this context. Although it is widely known that not Franklin but the molecular biologists James Watson, Francis Crick, and Maurice Wilkins received the Nobel Prize (at the time, Rosalind Franklin had already died of cancer), what is maybe less known is that, according to an assistant of Franklin, Watson and Crick gained access to Franklin's X-ray diffraction images and other data without her consent, using them for their own research (cf. Maddox 2003). Watson's disrespect against Franklin becomes clearly visible in his autobiography *The Double Helix* (1968) where he not only repeatedly calls Franklin 'Rosy' but also criticizes her for having a poor taste for clothing and for not wearing make-up.

"By choice she did not emphasize her feminine qualities. Though her features were strong, she was not unattractive and might have been quite stunning had she taken even a mild interest in clothes. This she did not. There was never lipstick to contrast with her straight black hair [...] it was quite easy to imagine her the product of an unsatisfied mother [...] The thought could not be avoided that the best home for a feminist was in another person's lab." (Watson 1968: 17-21)

Against the backdrop of the critique of the idea of a sole (male) inventor who develops and shapes a particular technology, many social construction of technology scholars foregrounded that specific socio-cultural and historical reasons (or even needs) are reason why technologies have their particular shapes and meanings. It always could have turned out differently. The development of technologies could have taken a different direction at any point in time. The "successful stages" of technologies are therefore "not the only possible ones" (Bijker et al. 1987: 22). Applying the principle of symmetry of the Strong Programme to technology led to the conclusion that: "technologies that work well are no different in this respect from those that fail" (Bijker/Law 1992: 3). In this regard, technologies that work and those that fail have to be analyzed in the same terms. Success and failure have to be regarded as the results of complex socio-technological interactions. Since social actors are the ones who give technical objects meaning, technology, analogously to knowledge, has to be considered as socially constructed.

Against this background, the social construction of technology (see Bijker/Pinch 1984; Bijker et al. 1987), alongside the social shaping of technology approaches (see MacKenzie/Wajcman 1985; Williams/Edge 1996)[3] stressed that technology is neither following its own momentum nor does it "emerge from the unfolding of a predetermined logic" (Williams/Edge 1996: 866). There is no such thing as an essence of technology. Without an essence, however, technical objects and systems cannot have only *one* possible consequence that would determinate social structures and relations (Sismondo 2004: 81; see also chapter two). As mentioned before, within this theoretical framework the meaning of technologies always comes from the social context in which they are embedded.

While technological determinism, in the end, can be reduced to the conclusion that we have to either adapt to the imperatives of technology and to technological change or get rid of modern technology altogether if we are to survive, social constructionist theories of technology shifted the focus to the relationship of technology and society, exploring *how* technologies are shaped by social factors, and how in turn technologies shape society. Since technology was understood as "shaped by and mirror[ing] the complex trade-offs that make up our societies" (Bijker/Law 1992: 3) the main argument that followed was that there is no such thing as "the 'purely' technological—that no such beast exists. Rather we are saying that the technological is social" (ibid: 4). The important point here is that the notion of the 'social' is not to be reduced on the notion of 'the sociological'. On the contrary, 'social' in this context means also always "political, economic, psychological—and indeed historical" (ibid.), as politics, economics, and much more "are thrown into the melting pot whenever an artifact is designed

---

3 In some cases Bruno Latour and Donna Haraway are also associated with this body of work. Adrian Mackenzie (2005: 383), for example, considers the works of Latour and Haraway as "some of the most well-known SCOT writings". While this might be true for Latour's earlier writings, in particular, for Latour's *Laboratory Life* (1979) and *Science in Action* (1987), Haraway's and Latour's later works are characterized by increasingly questioning key assumptions and conclusions of social constructionism. What is more, for both, *social* constructionism remains adhered to a specific kind of anthropocentrism and modernist thought. Consequently, both focus on questions of nonhuman agency and how the world is constantly enacted through material-semiotic practices. For this reason, I will not discuss Latour's and Haraway's writings in this section but will return to them later in this book by outlining a number of highly promising and productive thoughts that allow to understand technology and the body as always already entangled with one another.

or built" (Bijker/Law 1992: 3). Because technologies are never independent, never detached from society they cannot determine society or even history as an external force.[4] In fact, technology and society are folded into each other, forming a circular relationship. Society simultaneously shapes technology as technology shapes society. Both, society and technology thus have to be understood as different moments of the same phenomenon. The term technology does not refer to mere objects or systems detached from the social. The contrary is true, technologies are always embedded in and part of social practices. What is more, there is no longer an a priori distinction to be made between the social, the technological, the political, and the economic; rather, they are mutually constituting each other, fabricating what Thomas Hughes (1986) termed the "seamless web". Because technologies are inextricably bound up with human practices and interests there is also no such thing as value-neutral technologies. Since social interests, values, and relations shape technologies technology cannot be separated neatly from politics. "Even worse", as Bijker (2006: 682) puts it, "their 'definitions' are interdependent."[5]

THE TECHNOLOGICAL IS POLITICAL

The idea that technologies have political qualities can be traced back to Marx's work in which he considers technology as part of the productive forces. Marx had been read for a long time as technological determinist. And, it was not particularly hard to misunderstand Marx in this way, for example, when he argues in *The Poverty of Philosophy* (1920) [1847] that "[t]he hand-mill gives you society with the feudal lord; the steam-mill, society with the industrial capitalist",

---

4   Thomas Hughes' concept of *technological momentum* could be considered as an important exemption here. Hughes' main claim is that, even though most technologies start as socially shaped and controlled, they gather together more and more social actors as well as other technologies around them, growing to large technological systems. With time, technological systems acquire momentum and as a result "tend to exert a soft determinism on other systems, groups, and individuals in society" (Hughes 1987: 55).

5   However, in his more recent work Bijker (2006: 682) admits that "statements, such as 'all technology is political' or 'all politics is technological' may be true, but not very helpful". I shall ask later, how to understand technology as having political qualities, yet without grasping it as a mere place holder for 'the social' or as congealed social power relations.

implying that technological development changes social relations and consequently the very nature of society. Instead of craftsmen who were not bound to a specific place, the steam-mill demanded workers who were disciplined to be at a certain place – namely, the factory – for a certain period of time. By looking at industrialized societies of the late 19th century with their factories and smoking chimneys, with their engines, machines, and the thousands of poor and hungry workers gathering together around them, one might almost immediately come to the conclusion that while in "handicrafts and manufacture, the workman makes use of a tool; in the factory, the machine makes use of him" (Marx 1976 [1867]: 548). From such a point of view, the machine seems to become a strange power dictating the movements of the bodies of the workers. The worker, ultimately, seems to become the mere "living appendage" of the machine. In the so-called *Fragment on Machines*, Marx writes,

"The worker's activity, reduced to a mere abstraction of activity, is determined and regulated on all sides by the movement of the machinery, and not the opposite. The science which compels the inanimate limbs of the machinery, by their construction, to act purposefully, as an automaton, does not exist in the worker's consciousness, but rather acts upon him through the machine as an alien power, as the power of the machine itself." (Marx 1993 [1858]: 693)

It is against this backdrop that Langdon Winner (1977: 39) puts forward the argument that what Marx did was nothing less than to formulate "the first coherent theory of autonomous technology", that is, the idea that technology has gotten out of control.[6] In a similar way, Robert Heilbroner defends technological determinism at a time it became increasingly under pressure by turning to Marx's thoughts on the machine. Even though Heilbroner relativizes his undertaking by situating technological determinism into a specific historical era, "that of high capitalism and low socialism—in which the forces of technical change have been unleashed, but when the agencies for the control or guidance of technology are still rudimentary" (Heilbroner 1967: 345), he still defends the main claim of technological determinism, namely that technologies impose, if not dictate, certain social and political consequences upon society. What is more, Heilbroner

---

6　Winner mainly refers to a paragraph in *The German Ideology* where Marx writes, "This fixation of social activity, this consolidation of what we ourselves produce into a material power above us, growing out of our control, thwarting our expectations, bringing to naught our calculations, is one of the chief factors in historical development up till now." (Marx 1998 [1932]: 53)

also believes that "there is such a sequence—that the steam-mill follows the hand-mill not by chance but because it is the next 'stage' in a technological conquest of nature that follows one and only one grand avenue of advance" (ibid: 336). However, the claim that there is one – and only one – predetermined way technological development necessarily has to run-through, as if the development of technologies would follow nomological laws,[7] not only seems implausible but also cannot be derived from Marx's work on the capitalist development of the machine and large industry just like that. In fact, I think it would be a serious misunderstanding of Marx to think that when he talks of the machine he believes that technology would be the main or the determinant[8] productive force; or that the development of the productive forces has to be thought of as an autonomous process and consequently could be detached from the prevalent relations of production. It is also in this sense that Langdon Winner confuses technology as *part of* the productive forces with technology as *the only* productive force when he writes, "Man uses all of the instruments and tools—*the productive forces*—available to him in a conscious, productive manner." (Winner 1977: 37; italics JB) As a matter of fact, for Marx, the term productive forces not only comprise technology but also natural resources (animate and inanimate ones) and human labor as well as knowledge. Hence, for Marx, technology is only *a* productive force, even though an important one, but cannot be equated with *the* productive forces.

But if Marx was not a technological determinist why has his work been read for a long time as if he had argued for an understanding of technology as a powerful force determining society and even the course of history? The main reason for this problem might be found in the difficulty that technological determinism is an "elusive concept" (Bimber 1990: 333), characterizing an array of different

---

7   Heilbroner indeed understands technological development, at least to a certain degree, as a quasi-natural process. "If nature makes no sudden leaps", he writes, "neither, it would appear, does technology." (Heilbroner 1967: 337) For Heilbroner, technological development thus seems to be a linear, teleological process without any leaps, discontinuities, or setbacks. To be fair, it has to be said that Heilbroner nevertheless cannot be seen as a proponent of a strong technological determinism since he admits that to a certain degree the development of technology can be influenced socially. Yet again: only *influenced*.

8   It is important to understand that the German word *bestimmen* (to dictate) has not the exact same meaning as *determinieren* (to determine). If Marx perceives technology as a determinant force, it does not necessarily mean that he understands technology as *determining* society or even the course of history.

views about the role of technology in social change.[9] While stronger versions of technological determinism claim that there is an inevitable order of technological progress, that technology functions according to certain inherent laws, and that modern technology somehow got out of control, weaker versions of technological determinism emphasize that technology indeed is a dominant factor in social change, maybe even the most important one, but it is certainly not technology alone that changes the course of history.

It is true that, for Marx, technology, or more precisely, the machine, plays a crucial role in social change. But rather than understanding capitalist development only by looking at modern technology and its development, what Marx does instead is to take into account the relations and interactions between the development of the productive forces, the Bourgeoisie and its intentions, class struggles, and many other political, economic, and historical factors, tying them together. It is the *Fragment on Machines* in particular and the fifteenth chapter of the first volume of the *Capital* that resembles much of an actor-network-theory analysis or even a rhizomatic approach, when Marx describes the entanglements of organic and mechanical forces, human and nonhuman bodies as well as technical objects. Reading Marx diffractively through the philosophy of Gilles Deleuze and Bruno Latour[10] allows us to take in a perspective on Marx's work beyond narratives of the replacement of the organic by the mechanical or the fear that human bodies would become mere parts of the machine, yet without wiping away the political and ethical implications of these concerns.

If Marx describes the machine as "natural material transformed into organs of the human will" or as "the power of knowledge made into an object" (Marx 1971: 143), not only the distinction between material objects and knowledge becomes opaque but also the very boundaries between humans (or human bodies) and the machine do so as well. The machine seems to be more than a mere technical mechanism, an automaton, or a passive object. It could be said that for Marx the machine is a relation, a linkage or even an entanglement of human beings and nonhuman entities. This is, for example, the case when Marx writes that it is not the machine that is producing but always an association of workers and

---

9   Bruce Bimber (1990: 335) differentiates between three accounts of technological determinism: "Norm-Based Accounts", "Logical-Accounts", and "Unintended Consequences Accounts". For Bimber, "only Logical Sequence Accounts rightly deserve the label. While Norm-Based Accounts and Unintended Consequences Accounts may offer useful observations about technology and society, they are essentially neither technological nor deterministic in nature", he argues.

10 Cf. also Barla/Steinschaden (2012).

machines. At other places Marx discusses the machine not as a thing but as a method or a means for raising the relative surplus value (Marx 1976 [1867]: 492). In doing so, the machine becomes apparent as an assemblage of practices and knowledges, of mechanical, organic, and intellectual components or 'organs', in Marx's terms. What is more, the machine becomes visible as a mechanical-intellectual-social assemblage coordinating the diffused workers, linking together natural, artificial, and social 'components'. In particular, the chapter "Machinery and Large-Scale Industry" in the first volume of *The Capital* can be read as the philosophical attempt at describing networks populated by a multitude of humans and nonhumans. In his analysis of the factory and modern industry, Marx makes it clear that it is the machine that stabilizes Capitalism. Resembling Bruno Latour's (1991a) argument that social interactions alone are never enough to stabilize social relations and that it is only through material objects that society is made durable, Marx emphasizes that the whole period of manufacture is characterized by the lament of the Bourgeoisie about the indiscipline of the workers, until the machine comes in and fights this indiscipline by dictating the workers a rigid work schedule, linking them together with the machinery and its working rhythm. In line with Marx's argument, Friedrich Engels brings the example of the cotton mill to illustrate that in the factory everyone,

"men, women and children, are obliged to begin and finish their work at the hours fixed by the authority of the steam, which cares nothing for individual autonomy. The workers must, therefore, first come to an understanding on the hours of work; and these hours, once they are fixed, must be observed by all, without any exception [...] The automatic machinery of the big factory is much more despotic than the small capitalists who employ workers ever have been. At least with regard to the hours of work one may write upon the portals of these factories: *Lasciate ogni autonomia, voi che entrate!* (Leave, ye that enter in, all autonomy behind!)." (Engels 1978 [1872]: 731)

Reading Marx's work through the lens of actor-network-theory, the machine can be regarded as what stabilizes human beings as wage workers under capitalist conditions. It is not until the machine comes in that specific capitalist working conditions and hierarchies between capitalists and workers become fixed.

Precisely for this reason, the machine appears to be more of an actor, or even a network itself, than a mere object: Machines and material bodies (human and nonhuman ones alike) in their very entanglements not only constitute the factory but it is also the machine that is located in the middle of the resistance movement of the organizing working class. From such a perspective, it could even be said that to a certain degree it is the machine that marks out, or even creates, a durable working class by transforming a diffuse wandering mass of undisciplined workers into laborers; staying twelve and more hours in the factory subjected to a rigid control system for only starvation wages. It is not for nothing that Marx describes the machine as "the most powerful weapon" for suppressing revolts and strikes of the working class "which threatened to drive the infant factory system into crisis" (Marx 1976 [1867]: 562–563). And still, it would be a mistake to understand the machine itself as the universal culprit. Not only because there is no such thing as the machine *itself* cut loose from its entanglements with human and nonhuman bodies, the capital, social and economic relations, and many other things. Marx makes clear that Luddism and the swing riots[11] of the early 19th century were wrong to fight the machine itself since the problem lies not in the machine but in its "capitalist application" (Marx 1976 [1867]: 568). In fact, what Luddism only achieves would be that the capitalist would not have to "rack his brains any more, and in addition implies that his opponent is guilty of the stupidity of contending, not against the capitalist applications of machinery, but against machinery itself" (ibid: 569). However, it is again Engels who goes even further, criticizing that the struggle against the machine, in fact, should be a struggle *for* the machine; that is to say, the "machinery shall no longer work against but for them [the workers, JB]" (Engels 2005 [1845/1892]: 139). It has to be understood against this backdrop that Engels also criticizes anti-authoritarian Socialists and Anarchists such as Michail Bakunin for allegedly launching "a regular crusade against what they call the *principle of authority*" (Engels 1978 [1872]: 730). For Engels, modern industry depends on "combined action" and "whoever mentions combined action speaks of organization" (ibid: 731), meaning discipline and authority. Consequently, Engels argues that it would be stupid to think that authority would ever disappear. According to Engels, even in a truly socialist society, authority would not disappear but only express itself differently since it is a necessary culprit for organizing and

---

11 The Swing riots were a rural unrest of agricultural workers in the early 1830s in the southern counties of England. Similar as in the case of Luddism, the riots adopted their name from the threatening letters sent out to manufacturers and farmers signed with the fictional name Captain Swing. However, it was not the case that the workers turned their rage solely against new machines, or, more precisely the threshing machine, as Marx portrayed it rather one-dimensionally. Indeed the workers were also furious because of extremely low wages, poor living and working conditions, and not least because of years of bad harvest (what culminated ten years later in Ireland in the Great Famine and mass exodus after parts of the potato harvest spared from the potato blight had been shipped to England). Cf. also Griffin (2012).

governing large numbers of people, as well as the methods of production, under specific material conditions. Thus, Engels makes clear that the wish "to abolish authority in large-scale industry is tantamount to wanting to abolish industry itself, to destroy the power loom in order to return to the spinning wheel" (ibid.).

By taking in such a perspective on Marx's work, the machine appears almost as a subject rather than a mere inanimate mechanical object. One could go even further and say that, in Marx's work, it is neither the machine nor the workers, and even less the capitalist who produces, but rather an assemblage consisting of material and immaterial, organic and mechanical entities, as well as of political, social, and economic relations. Humans and machines, human bodies and material technologies, in their entanglement with one another do not form contrasts. From such a perspective, at least for the example of the factory, human bodies and machines can no longer be thought of in categories such as naturalness and bodily separability or closure. Similarly, it becomes difficult to differentiate between autonomous purposeful action and (technological or economic) determination. The idea of the machine as an automaton and object experiences a reconfiguration toward a notion of the machine as a complex, opaque assemblage consisting of a number of heterogeneous entities which are entangled with each other, making it hard to determine where material bodies end and technologies begin. This is particularly the case when Marx describes the machine as an assemblage that "consists of a number of mechanical and intellectual organs, so that the workers themselves can be no more than the conscious limbs of the automaton" (Marx 1971: 132), and argues further that the machine "is itself the virtuoso, with a spirit of its own". And "just as the worker consumes food, so the machine consumes coal, oil, etc. (instrumental material), for its own constant self-propulsion" (ibid: 133). It is precisely in this sense that Marx's analysis of the machine is not only an inherently political one – by bringing together a multitude of heterogeneous entities such as workers, capitalists, technical objects, factories, as well as political discourses, parliamentary debates, statutes, debates about public health and illness, economic statistics, strikes, riots and acts of sabotage, and many other things, but also tells us much about the entanglement of material bodies and technologies in the factories of the late 19th century.

Even though Marx's thoughts on the machine can be read as one of the earliest political philosophies of technology, it was Langdon Winner who stressed the political character of technical objects more lucidly, arguing that social power relations manifest in them establishing social order. Trying to provide "an antidote to naive technological determinism" (Winner 1986: 21), Winner identifies two ways in which technical objects can contain political qualities. First, technologies can be what Winner calls "inherently political" (ibid: 20). Inherently political technologies are political in themselves and "appear to require or to be strongly compatible with particular kinds of political relationships" (ibid: 22). The adoption of such technologies would unavoidably link to "particular institutionalized patterns of power and authority" (ibid: 38). Nuclear weapons, for example, but also large production factories, as the ones described by Engels, which demand the adaption or even the subordination of the workers to the machine designate, according to Winner, inherently political technologies. The justification for authority, in these cases, would derive from the technical objects themselves. One can call this the strong version of the idea that technical objects have politics. While this idea makes perfect sense for describing obviously undemocratic technologies such as nuclear power plants or ballistic missiles which rely on centers of power with restricted access, it also resembles technological determinist theories by implying that some technologies would determine unavoidably the very ways people live together as social beings, leaving no flexibility for interpretation and reinterpretation.

However, for Winner, technical objects can also be political in the way that they "embody specific forms of power and authority" (ibid: 19). Referring to the hanging bridges over the parkways on Long Island, built by Robert Moses between the 1930s and the 1960s, Winner points out how racism and political interests manifest themselves in technical objects. As the bridges were built remarkably low only cars but not buses could pass under them denying the residents of New York City dependent on public transit, and thus particularly the poor and people of color access to the beach. Since material technology embodies social power relations and practices, building bridges, for example, can, according to Winner, function as a kind of social engineering. Thus, within the weak version of the idea that technology has political qualities technologies can be regarded as ways of "building order in our world" (ibid: 28).

But, despite the fact that there is no doubt that technologies have political qualities it seems less convincing to understand technology *primarily* as materialized social power relations. Moreover, Winner's claim that the low hanging bridges are "there for a reason", that is, to "achieve a particular social effect" which would be to limit access to the beach and with that keep the poor and people of color "off the roads" (ibid: 23), implies that there would be a guaranty that the interests and aims inscribed in technologies could always be achieved. The story behind the low hanging overpasses on Long Island can also be told differently. Bernward Joerges (1999), for example,[12] offers an alternative version of

---

12 Cf. Woolgar and Cooper's (1999: 436) objection that "Moses' bridges could not and did not obstruct the buses", as not only timetables for buses would show but also

the story by stressing that Robert Moses had no racist intentions in mind when he was planning the bridges. Rather, Moses had only followed the emerging trend of creating "environments for the automobile society and what it represented" (Joerges 1999: 420) in the United States of the late 1940s. But, to be fair, also in this version of the story it is hard to deny that this happened at the expense of the poor residents of New York City, as for example the Bronx got not only cut off from recreation areas but also divided by a number of expressways.

Winner's assumption that social actors and groups inscribe specific interests into technologies with the aim to sediment them in artifacts for the pursuit of power and dominance not only runs the risk of understanding technology first and foremost as a negative force but also reduces 'the political' to "arrangements of power and authority in human associations as well as the activities that take place within those arrangements" (Winner 1986: 22). With that definition it ignores the possibility that technology as well as politics can also be the source for emancipatory actions. What is more, Winner defines technology as "all of modern practical artifice" (ibid.). Taking his examples into the focus, however, technology is reduced to the exploration of *material* artifacts. Yet Winner is not alone with this problem. Despite the fact that it has been emphasized by many SCOT scholars that, "technology is more than a set of objects or artefacts" (Wajcman 1991: 149) as it always also involves knowledge, beliefs, and practices, technology is still mostly theorized as artifacts or systems.[13] That being so, the idea of the social construction of technology, in many cases, remains synonymous with the social construction of technical artifacts.

Drawing on insights of this body of work, feminist scholars have argued that precisely because technology is political, it is also gendered. In doing so, feminist scholars not only highlighted that industrial, military, and commercial technologies are masculine both in a symbolic and in a very material sense, but also located the connection between gender and technology historically in the development of industrial capitalism in the middle of the 19th century. Cynthia Cockburn, for example, argued that the exclusion of women from production processes within the Industrial Revolution helped to establish a mode of technology that became "one of the formative processes of men" (Cockburn 1985: 54). Similarly, also Judy Wajcman (1991: 21) identified "the foundations of male dominance of technology" in the gendered division of labor. Alongside many other scholars within this tradition, Wajcman consequently argues that

---

residents of New York City stated, who have been using public transportation over the last decades to get to the beach and to other places.
13 Cf. also Bijker et al. (1987: 4) and MacKenzie/Wajcman (1985: 3-4).

technology became an integral part of masculine gender identity because "the machinery was designed by men with men in mind, either by the capitalist inventor or by skilled craftsmen. Industrial technology from its origins thus reflects male power as well as capitalist domination" (ibid.). However, the exclusion of women from the production process, or at least their usurpation into low-paid jobs, was perhaps not so much a direct effect of industrial capitalism and the machinery (at least not alone) but ironically also the outcome of the beginning organization and foundation of labor unions which not only denied membership to women but often also actively opposed the employment of skilled female workers in factories. Furthermore, a series of Acts (later known under the term "Factory Acts") that were passed by the Parliament of the United Kingdom between the 1830s and 1850s not only regulated the working hours for women and young persons to a maximum of ten hours a day but also lead to mass layoffs of women, who at that time made up 40 percent of the factory workers.[14] What becomes visible here is that class struggle cannot be understood properly if only class is taken into account. Instead of focusing on monocausal explanations, the intersections of different social categories such as gender, race, dis/ability, and many others (and in so doing the multiple axes of difference and inequality) have to be taken into account critically.

Employing careful historical analyses, feminist scholars argued that technology and (hegemonic) masculinity are to be understood as closely linked to each other. It was this connection between technology and masculinity that had a lasting effect on the picture of technology in society. Not only that technology had been reduced to machines for a long time but also factory work and the operation of machines had been regarded as a male activity with the result of erasing the fact that historically women were not only operators but also inventors of a number of machines and technologies. As a matter of fact, many of the first computer engineers and programmers were women. For example, the programmers of the first general-purpose electronic digital computer ENIAC (Electronic Numerical Integrator and Computer), a computer developed by the US Army for the purpose of calculating ballistic trajectories during the 1940s and 1950s. Or the programmers Grace Hopper, Gertrude Tierney, and Jean E. Sammet who developed the programming language COBOL (Common Business-Oriented Language) in the late 1950s. Following the main claim of the social construction and social shaping of technology framework, namely, that technical objects and systems are never only 'technological' but always also political, economic, and socio-cultural, scholars within this tradition consequently argued that gender

---

14 I would like to thank Sam Wade for this helpful remark.

hierarchies and identities play a crucial role in the design and application of technologies. Furthermore, since social relations, and in particular, gender hierarchies shape technical artifacts and systems technology has to be regarded as artifacts incorporating "masculine values" (see Cockburn 1985; Wajcman 1991; Faulkner 2001; Faulkner/Lohan 2004).[15]

The idea that the "production and the use of technology are shaped by male power and interests" with the effect that technology has to be seen as "imprinted with patriarchal design" and therefore as a kind of "masculine culture" (Wajcman 1991: 162-163) soon gave way to a social constructionist understanding of gender (identities) and technology as co-produced or co-constituted. Translating Langdon Winner's rhetorical question 'Do artifacts have politics?' into the question 'Do artifacts have gender?' (Berg/Lie 1995), the argument was brought to the fore that because technologies are shaped by gender norms and hierarchies as much as gender relations, in turn, are shaped by particular technologies, technology and gender have to be regarded as deeply intertwined. Gender identities and relations are not only 'inscribed' into technical artifacts or embodied by technologies but also 'materialized' in them (Faulkner/Lohan 2004: 83; Wajcman 2010: 149). Technologies, thus, have to be understood as materialized and condensed social (power) relations, meaning not only gender hierarchies but also racial inequalities.

Against this backdrop, feminist scholars within this constructionist framework also highlighted a crucial blind spot of mainstream social construction of technology approaches. As pointed out earlier, a central claim of the social construction of technology paradigm is that, following the principle of symmetry of the Edinburgh School, successful technologies were successful not because they were 'better' than less successful technologies but rather because they were accepted by a larger number of relevant social groups. However, it is precisely the concept of the relevant social group that has to be problematized for its

---

15 It is against this particular backdrop that Judy Wajcman (1991: 126) argues that modern transportation technologies are not only gendered but also "that the transport system, and in particular the dominance of the car, restricts women's mobility and exacerbates women's confinement to the home and the immediate locality". While it is certainly true that transportation technologies are gendered, it cannot be denied that technologies such as the bicycle, the train, or the automobile had also contributed to, or even allowed in the first place a higher mobility of women as well as poorer people in general. The concern of men, the attempts of patronizing and trying to refuse women access to these technologies in the more recent past illustrate this point clearly. Cf. on this argument also Bijker et al. (1987: 34).

gender-blindness. The reality of gender hierarchies and the question of how certain assumptions about gender are inscribed into technologies had been largely ignored. What followed from this was the fallacy that "as long as *women* do not appear as important actors or as a relevant social group, *gender* is not a relevant category" (Berg/Lie 1995: 344). The problem, however, lies even deeper, as it could be argued that the concept of relevant social groups itself is a problematic one as it ignores the important fact that there are crucial asymmetries in power between different social groups. There are a lot of examples that women and people of color raised their voice for or against a particular innovation but did not get enough attention. Does this mean that women, ethnic and economic minorities, and other social groups were not *relevant* because they were ignored? Does this mean that race, sex/gender, and other categories of inequality are not important in critical analyses of innovation and technological development? Mainstream approaches of a social construction or social shaping of technology run the risk of erasing, or at least ignoring, the interventions of marginalized groups in processes of the design, development, and adoption of certain technologies.[16]

Following the premise of social construction and social shaping of technology approaches that technology cannot be separated from politics and that technical objects have to be regarded as artifacts embodying social power relations and interests, feminist scholars in the tradition of this body of work have argued that technology *is* "social relations" (Cockburn/Dilic 1994: 8). In *The De-Scription of Technical Objects* (1992) Madeline Akrich argues that engineers, inventors, designers, and other decision makers 'inscribe' their values and "vision of (or prediction about) the world in the technical content of the new object", defining the end product of this work "a 'script' or a 'scenario'" (Akrich 1992: 208). A script, thus, can be understood as a program containing specific rules and patterns of behavior, aiming at pushing social actors "into specific roles" (ibid: 215). It is against this background that for Akrich "to say that technical objects have political strength" makes perfect sense since technologies "may change social relations, but they also stabilize, naturalize, depoliticize, and translate these into other media" (ibid: 222). In doing so, Akrich problematizes that "social constructivism denies the obduracy of objects and assumes that only people can have the status of actors" (ibid).

---

16 In recent years there have been some attempts to incorporate feminist critique into the social construction of technology framework. See, for example, Oudshoorn/Pinch (2003: 4).

In much of the social constructionist literature on technology, technologies represent the result of "a series of specific decisions made by particular groups of people in particular places at particular times *for their own purposes*" (Wajcman 1991: 162; italics JB). What is more, not only have technologies been considered primarily as social relations or materialized and condensed social or individual interests, but it also has been argued that "those who design new technologies are, by the same stroke, designing society" (Faulkner/Lohan 2004: 322). Even though this argument has been softened sometimes by following the main assumption of the social construction of technology paradigm, namely that "the process of designing technologies and societies is not straightforward, because technology is subject to considerable interpretative flexibility" (Faulkner/Lohan 2004: 322), and that "designers and promoters of a technology cannot *completely* predict or control its final use" (Wajcman 1991: 104; italics JB), the claim is still left hanging in the air that there is a good chance that technologies would not only embody social relations and particular interests but also do function according to particular goals and interests that had been inscribed into them.

While it is plausible to consider gender as a social relation, it seems already less convincing to understand technology, too, first and foremost as condensed social power relations. Certainly, there is no doubt that technologies are political. One could even argue from a Latourian perspective that technologies are what lend social actions and norms a certain stability, and that technology, therefore, is nothing less than "society made durable" (1991a).[17] However, considering technology first and foremost as congealed and condensed social power relations not only runs the risk of equating technology with material artifacts and systems, and in doing so ignoring that technology also denotes a particular mode of knowing and being, but, in assuming that specific intentions could be inscribed into technologies, such a perspective also comes critically near to an understanding of technology as mere politics by other means or even as but a "lust of power", as Bruno Latour (2005a: 85) polemically puts it. What is more, in many approaches within this framework technology resembles much of a fixed architecture. For example, if Robert Moses's bridges are, as Langdon Winner emphasizes, really 'there for a reason', it is implied that once built, the bridges will unfold ceaselessly the individual or collective biases, interests, and imperatives inscribed into them – at least until they are torn down or substantially converted. But from such a point of view technology once again appears as a determining force.

---

17 I shall return to a detailed discussion of Bruno Latour's philosophy of technology in the second chapter of this book.

TECHNOLOGY AND THE BODY: AN ASYMMETRICAL RELATION

Along with the understanding of technology as the materialization of social power relations however goes another difficulty that many of the approaches discussed above share. While technological determinist theories were often pervaded by cultural pessimism and tended to be fatalistic in their explanations they also had something to say about material and dynamic bodies and their entanglements with technologies, even though, they often draw questionable conclusions from it. Thorstein Veblen (1914: 307), for example, argued that it is, in fact, the machine that uses the workers and their bodies and not the other way around. Lewis Mumford (1967) developed the figure of the 'megamachine' to describe what he understood as superstructures of disciplined and regulated human bodies and material technologies. Martin Heidegger (1979: 99) holds that we do not "have" a body, rather "we 'are' bodily". "Being-in-the-world" (*Dasein*), thus, was always bodily being-in-the-world for Heidegger. And it was precisely this specific bodily mode of being that Heidegger saw threatened by modern technology and instrumental rationality that would degrade the body and dissect it into 'parts', transforming it into a kind of standing reserve (*Bestand*). Turning to the example of the typewriter Heidegger consequently argued that 'modern' technology would not only degrade the hands in the way it "tears writing from the essential realm of the hand" and in doing so "makes everyone look the same" (Heidegger 1992 [1942/43]: 81), but would also destroy communication itself. This pessimistic view on modern technology resonates with Günther Anders's (1956; 1980) idea of 'the outdatedness' of the human. For Anders, not only our bodies but also our humanity would be submitted to modern technologies. Technology becomes our destiny. In a different and yet similar fashion Jürgen Habermas (1970) understood modern science and technology as 'ideology' and instrumental reason. It is exactly this idea that finds its contemporary continuation in Gernot Böhme's (2012: 215) thesis of an invasive technification of the natural body and with that a dissolution of "the nature that we ourselves are".[18]

Social constructionist theories of technology primarily focused on the social construction of technical objects (see Pinch/Bijker 1984 for bicycles; Law/Callon 1992 for the social construction of aircrafts; Misa 1992 even for steel; and MacKenzie 1998 for nuclear missiles), the coproduction of technology

---

18 I shall return to a more in-depth problematization of some of these theorizations of technology and the body in chapter three.

and society (see Bijker et al. 1987; Hughes 2005), or more recently user-technology relationships (see Oudshoorn/Pinch 2003). But, material bodies in their multiple and dynamic entanglements with technologies had only been rarely taken into account. While approaches within this body of work broke with the dichotomy of technology and society by moving "constantly between the technical and the social" (Akrich 1992: 206) they left largely untouched the dichotomy of technology and the body, and with it the dichotomy of the natural and the artificial, the organic and the mechanical.

Feminist scholars were not only successful in highlighting the blind spots of mainstream social construction of technology analyses. More importantly, feminist scholars within the tradition of this body of work were also the ones who shifted the focus toward the material body. This was no coincidence, considering the long standing link that had been established between the female body and 'Nature', resulting in the fact that for a long time women, unlike men, had not been allowed to jettison the (biological) body in order to become the modernist, disembodied (male) Cartesian subject with its view from nowhere. Emerging out of the protests of 1968, women's liberation groups began to spread in the early 1970s all across the United States and Europe. Grassroots activism and academic critique went hand in hand in debunking patriarchal beliefs and social structures. Employing slogans such as "our bodies, ourselves",[19] the women's health movement demanded the right of autonomy for women and with it the right to decide over the own body. Simultaneously, biology and medicine have gained center stage as politically contested grounds. Not only had prevailing beliefs about the female reproductive system and with it the alleged nature of women been contested, but also the pathologization of female bodies was more and more criticized as a violent form of oppression. In particular, biological and medical theories which claimed that there is a direct effect of physical and physiological processes on individual behavior, were criticized for not only being androcentric by understanding the male body as the unmarked norm, while female bodies were regarded as the other, meaning, as the deviation from that norm, but also

---

19 *Our Bodies, Ourselves* was the name of a book published by the Boston Women's Health Book Collective in 1973 focusing on the, at that time, still tabooed issues of women's sexuality and health. The conservative Intercollegiate Studies Institute with its aim to fight alleged political correctness and liberalism still keeps the book in its list of the "50 Worst Books of the Twentieth Century", where it is in company with other allegedly 'bad books' such as Herbert Marcuse's *One-Dimensional Man* (described as "the Bible of the sixties and early postmodernism"), John Dewey's *Democracy and Education*, and Karl Popper's *The Open Society and its Enemies*).

for considering female bodies as not equivalent (meaning, not of equal value) to male ones with regard to their material, affective, emotional, and cognitive capacities. This was the case with many biological evolutionary theories and behavioral theories that reduced human action, thinking, and feeling one-dimensionally to biological, genetic, and hormonal aspects. Edward O. Wilson, for example, argued in *On Human Nature* (1978: 125) that it would be a "biological principle" that males are "aggressive, hasty, fickle and undiscriminating" while females would be more passive and "hold back until they can identify males with the best genes". In a similar way, more recently, the biologist Randy Thornhill and the anthropologist Craig Palmer (2000) were aiming for a self-proclaimed "ideology-free corrective" to feminist and other critical analyses[20] that understand rape as not (primarily) sexually motivated. Thornhill and Palmer argue that rape might be an "evolutionary adaption", meaning a so-called 'reproductive strategy' to increase the chance for offspring. In doing so, Thornhill and Palmer not only enmesh rape with sexual reproduction, bringing with it the heteronormative fallacy which ignores that survivors of rape are not only women (or, more precisely, "females of reproductive age", in Thornhill and Palmer's terms), but also ignores questions of power and dominance, institutionalized and structural violence, social role models, as well as phenomena such as war and prison rape, or so-called 'corrective rape'.

By identifying many alleged laws of nature as, in fact, the Laws of the Father (Lacan), feminist scholars pointed out how these theories were used to legitimate

---

20 While Thornhill and Palmer remain rather reserved with obvious sexist assumptions and claims, such as, for example, that women's dress choice has to be seen as a "sexual strategy" that would signal "sexual availability" (Thornhill/Palmer 2000: 182-183), implying that females therefore should rethink the ways they dress as well as their appearance and behavior around males if they do not want to get sexually harassed, the back cover of the book much more explicitly illustrates the anti-feminist cadence and the animosity against social theories and the humanities in general lying at the heart of this project by stating that the book "will force many intellectuals to decide which they value more: *established dogma and ideology*, or *the welfare of real women in the real world*. [...] By dispelling the most dangerous social science myths attempting to explain rape [that is, that rape is not primarily sexually motivated, JB] Randy Thornhill and Craig Palmer have *scientifically unveiled* one of the most dangerous scourges of the human condition for *what it really is* [...] Rape can cost males very little and females very much. That's why men do it, and women don't. It took a horde of humanists to obscure that fact. It took a biologist and an anthropologist—Thornhill and Palmer—to clear it up" (italics, JB).

and reinforce social power relations and dominance, and with that the social and economic positions of women but also of people of color by referring to 'Nature' and 'naturalness'. What followed was that the use of biology, or more specifically *biological theories*, had been debunked as a strategy to rationalize and legitimize social inequalities. Moreover, it had been pointed out that such theories present a highly problematic picture of human behavior and human nature, exhibiting sexist, racist, and anthropocentric values. Consequently, feminist scholars and activists have argued that neither gender nor, for that matter, race are rooted in biology and nature. Rather, both represent social constructions, meaning specific socio-historical concepts and categories of inequality. Against this backdrop, the argument was that not only technology and science but also the material body has to be regarded as a social construct.

Feminist critique emphasizing the socially constructed character of phenomena such as nature, the gendered body, and much more, has been equally a powerful antidote to technological determinism and to biological determinism and essentialism. But while the body has been explored in many productive ways, when it came to understandings of the relationship of technology and the body, and with that to the question of what kind of conjunctions connect them, the latter often remained a mute, disciplined, disempowered, and technologically molded object in much of the constructionist body of work. This is especially, although not solely, the case for contestations of 'new' reproductive technologies, which, as a result of the mentioned long lasting association of female bodies with nature, have been (and still continue to be) a necessary subject of critique for feminist interventions. A central argument that is often brought up here is that pregnancy today is controlled by ever more sophisticated and intrusive technologies. Consequently, reproductive technologies are understood as "means for exercising power relations on the flesh of the female body" (Balsamo 1996: 82), with the effect that female bodies are treated as if "they were all potential maternal bodies, and maternal bodies as if they were all potentially criminal" (ibid: 83). However, from such a point of view not only does the body become primarily a technologically disciplined, surveilled, and subjugated body but also technology becomes something external to the body, a mere means to an end, aiming at the manipulation and with it the domination of the (here especially female) body. It is true that pregnancy today cannot be separated from technological apparatuses and instruments but neither can it be from political, cultural, juridical, religious, and economic practices and norms which, as a matter of fact, reach well beyond the actual birth of the child. Therefore, the question is, does such an account of technology and material bodies, precisely with its critique of a technification of the body, then not imply that something that was once non-technological now has become 'technologized'? Furthermore, does such a perspective, then, not run the risk of thinking again in terms of the dichotomy of the natural and the artificial, the active and the passive, as well as mere inert matter on one side and powerful discourses and technologies, understood as condensed social power relations, breathing life into or bringing death to material bodies, on the other?

In *Technologies of the Gendered Body*, Anne Balsamo (1996: 23) aims at avoiding "essentialist and anti-essentialist perspectives" on technology and the body. While Balsamo acknowledges that "the body is at once both a cultural construction and a material fact of human life" (ibid: 33), in an unsettling way, this body still remains a silent and passive one in her account. Most importantly, it remains a body that is constantly threatened by both new biomedical technologies and by technologies of the self. It is in this sense that Balsamo argues already at the very beginning of her book that what we would witness today is a dramatic "technological refashioning of the 'natural' human body" (ibid: 9), that is, the transformation of (especially female) bodies into normalized objects through modern technologies as well as through a specific medicalized gaze,[21] which is always "a technologized view" (ibid: 57). It appears that for Balsamo the body represents first and foremost a target for modern technologies (including technologies of the self) and powerful discourses. Thus, Balsamo not only diagnoses a technological transformation of "the natural body [...] into a sign of culture" (Balsamo 1995: 225) but also a "technological fragmentation of the body" which would function "in a similar way to its medical fragmentation: body parts are objectified and invested with cultural significance" (ibid: 234). The body here not only tends to become a mere object upon which social power acts, disciplining it and transforming it, but is also reduced to its mere surface, that is inscribed with culturally specific practices and subject to technological interventions.[22] The question here is, does such a perspective on technology and the body result from a somewhat one-dimensional reading of Foucault's concept of technologies of the self and how they relate to material bodies, or, can it, at least in parts, already be found in Foucault's philosophy itself?

---

21 Balsamo (1996: 57) terms this specific technologized gaze, following a concept of the feminist media theorist Mary Ann Doane, "the clinical eye".

22 See also Jana Sawicki (1999) for a very similar account. Sawicki uses Foucault's concept of bio-power to problematize acts of disciplining (female) bodies and processes of making them available through new bio- and reproductive technologies, but in doing so also limits herself largely to an understanding of (female) bodies as the mere end-product of technologies of the self and powerful discourses.

When Foucault described the advent of a new source of power, namely, disciplinary power, in the 18th century, it was, in particular, the body that he had in mind as the direct locus of social control; that is, as the principle target of disciplinary power. In Foucault's understanding disciplinary power denotes something that is "addressed to the body, to life, to what causes it to proliferate, to what reinforces the species, its stamina, its ability to dominate, or its capacity for being used" (Foucault 1978: 147). Disciplined by the microphysics of power the body is subject to a vast number of small transformations, adapting it to the needs of the political economy. Through relentless observation, control, and discipline, rather than through the application of excessive force, the body is eventually molded into a 'docile body'. Foucault, however, was not only interested in the ways disciplinary power acts upon the individual and the collective but also in the question how the individual also acts upon herself and her body. This acting upon oneself is what Foucault coined as technologies of the self, as "matrices of practical reason" which "permit individuals to effect by their own means, or with the help of others, a certain number of operations on their own bodies and souls, thoughts, conduct, and way of being, so as to transform themselves" (Foucault 1988: 18). Technologies of the self, thus, function according to Foucault as a strategy through which individuals ensure themselves to be in control over their bodies. In this sense, technologies of the self are not only repressive but can also be generative, that is, they might open up new possibilities through cultivating new (bodily) capacities. For the body, however, this means that it is not only targeted and produced by biopower and disciplinary power, which "invest it, mark it, train it, torture it, force it to carry out tasks, to perform ceremonies, to emit signs" (Foucault 1977: 25), but also by technologies of the self. It is in this very sense that, on the one hand, Foucault urges us to take into account the body as well as the practices through which it is marked, shaped, forced to carry out tasks, and disciplined. Foucault even wonders "whether, before one poses the question of ideology, it wouldn't be more materialist to study first the question of the body and the effects of power on it" (Foucault 1980: 58). On the other hand, however, *this* body appears only to be a passive site for understanding the workings of power: the inscription of social norms and laws, ideologies, technologies, and other practices which are exercised on the body from an 'outside'. In fact, Foucault tells us surprisingly little about the capacities of bodies to be unruly and to kick back. The body Foucault describes through many figures and metaphors as a set of heterogeneous, multidirectional forces remains a body under siege, a body constantly molded and reconfigured by cultural, economic, political, juridical, and technological forces. It is only in a late interview where Foucault emphasizes that power might also produce "responding claims and affirmations", exposing power after its investment in the body to a degree that it could even be spoken of a "counter-attack" (ibid.) against technologies (once again, including technologies of the self). However, Foucault does not go into detail here. In the end, the body remains a surface upon which culture is imprinted in Foucault's philosophy. It is precisely this reducing of the body on its mere surface what ignores the 'inside' of bodies, that is to say, the "reproductive organs, lungs and heart, glands and capillaries", as Kathy Davis (2007: 54) emphasizes, and thus fails to answer the question of what bodies are capable of.

While it is true that there is no such thing as a natural body, that is, a body outside of its cultural significations, the assumption that the body beyond its social and discursive construction remains forever hidden (and with it also its non-discursive parts) takes every possibility to analytically grasp material bodies in their liveliness and unruliness. Therefore, such a perspective runs the risk of understanding material bodies again only as passive receptors for social control. The body remains a passive and blank transcendental body, waiting to be inscribed by powerful discourses and reshaped by invasive technologies. Furthermore, the body not only remains a singular, clearly definable body but also the idea is evoked that rather than having and being a body is understood as always inseparable from one another, the body becomes a mere object that is owned by the autonomous subject who, especially with the help of technology, is always in charge.[23] The head of the king has been cut off but the autonomous subject still remains the monarch, residing 'within' the palace of the body, as it is visible,

---

23 I am not saying here that it is not useful or sometimes even necessary for ethical and political reasons to emphasize that one is the owner of their own body, especially when it comes to sexual and reproductive freedom, abortion, hospitalization, incarceration, or acts of resistance against the economic commercialization of the body, its forces, parts, activities, and 'products'. Rather, my point here is that such a line of argument that renders the body first and foremost a property of the self not only reinforces the Cartesian dualism of subject and object, of active mind and passive matter, but also ignores that, as a matter of fact, the subject is never in full control over her own body. While it is possible, for example, to alter the surface of the body (meaning, the skin or certain bodily shapes) and to a certain degree also such things as, for example, the speed of the respiration with the help of certain technologies, drugs, and techniques, it is almost impossible to, say, let the heart or any other organ stop functioning willingly or to control the speed of the reproduction of body cells, and so on.

especially, in large parts of the debate on the critique of the neoliberal optimization and commodification of the body.²⁴

What follows is that while matter and material bodies are conceptualized first and foremost as social constructions, there still remains a nature 'behind' these constructions in much of the literature discussed above. With that the existence of a fixed matter and an essential natural body is assumed tacitly. Of course, there is no direct access to this nature, which remains hidden behind its social construction, since a vast gap stretches out between the realm of the material on the one side and their cultural interpretation and representation on the other. However, once this implicit positing becomes foregrounded, once the belief that matter and material bodies are only blank slates for social and cultural inscriptions is debunked as hardly tenable, the idea of the social construction of the body loses much of its plausibility.

It is precisely this social constructionist understanding of the material body as a passive, molded object that threatens to melt away under the powers of technology and technological rationality, that has come under critique. While there is no doubt that social constructionist feminist approaches were highly successful in deconstructing biological determinist and essentialist discourses that naturalized the (female) body, they also have been, more recently, problematized for ignoring the fact that bodies have their own forces too, and are more than just inert matter. Consequently, social constructionist and poststructuralist feminist theories²⁵ have been criticized for tending to replace biological and technological

---

24 See, for example, Nancy Scheper-Hughes and Loïc Wacquant (2002) for a problematization of the commodification of the body and its 'parts', and Philip Vannini and Denis Waskul (2006) for an understanding of bodies as "traces of culture". While these approaches deliver valuable insights and critique, the body itself all too often seems to remain a mute object on that culture, discourse, and technology acts upon, fundamentally reshaping or even 'denaturalizing' it.

25 Sometimes social constructionism and post-structuralism are equated even though many poststructuralist scholars have in fact problematized *social* constructionism. Judith Butler's *Bodies That Matter* (1993), for example, can be read as a powerful critique on prevailing social constructionist ideas of the body. If Butler rethinks the body as constructed she does not mean solely *socially* constructed. Rather, what Butler does is precisely to rethink the very meaning of 'construction' itself. It is against this backdrop that Butler highlights that "[t]he discourse of 'construction' that has for the most part circulated in feminist theory is perhaps not quite adequate" (Butler 1993: xi). Butler's position becomes even more clear when she argues that the notion of social construction "implies a culture or an agency of the social which acts upon a nature,

determinism with social or cultural determinism and hence for being 'anti-biology'. Elizabeth Wilson, for example, argues that although "questions of 'the body' have become increasingly fashionable in all manner of feminist projects [...], the schedule of feminism's anti-biologism has been little altered" (Wilson 1998: 14-15). Similarly, Noela Davis problematizes the ways in which biology and material bodies are conceptualized in feminist social constructionist literature, arguing that inherent to approaches within this body of work "is the conventional assumption that the biological and the social are two separate and discrete systems that then *somehow interact*" (Davis 2009: 67; italics JB). While for Elizabeth Grosz (2004: 2) it is even "the nature, the ontology, of the body" that "we" would have "forgotten" in the wake of social constructionist theories of biology and material bodies.

It is against this backdrop that Sara Ahmed identifies the accusation that constructionist feminisms are "being routinely anti-biological" (Ahmed 2008: 24) and that many feminist scholars within this body of work would have ignored matter and material bodies as a specific "founding gesture" of new materialist theories. Although not constituting a homogenous body of work, 'new materialisms' refer to approaches influenced by feminist materialist theories, poststructuralist theory, feminist epistemologies, and philosophies of science and technology.[26] Going beyond essentialism and constructionism as well as materialism and poststructuralism, new materialisms bring together situated epistemologies and relational ontologies. Instead of assuming a gap that separates the social world from the natural or material world, only to reconnect them through representation in a second step, remaining in the frameset of classic philosophy of

---

which is itself presupposed as a passive surface, outside the social and yet its necessary counterpart" (ibid: 4). However, when it comes to her notion of materialization Butler focuses too narrowly on the question how "discourse produces the effects that it names" (ibid: 2), not only leaving out the question of how matter comes to matter but also understanding materialization predominantly as the effect of discourses. I shall return to this argument in chapter three.

26 Iris van der Tuin traces the term 'new materialism' back to the work of Rosi Braidotti and Manuel DeLanda, both drawing on Gilles Deleuze and Félix Guattari's philosophy (See Dolphijn/van der Tuin 2012: 48). The philosophy of Friedrich Nietzsche and the work of Henri Bergson mark also important genealogical points of reference. At the 9th Annual New Materialism Conference at Utrecht University, Rosi Braidotti and Arun Saldanha stressed in their keynotes that the emergence of new materialist thinking has to be closely linked to the question of how to do politics differently after the failure of the protests of 1968.

consciousness and transcendental idealism, new materialist theories discard the idea of a res extensa as a passive realm that only awaits the power of the mind to be set in motion and to be 'brought to speak'. Instead, the question for the potentiality of matter and the material world is brought back in and critically renegotiated – this time, however, matter is considered as dynamic and agentic. In doing so, new materialisms provide a common ground for different disciplinary concepts and methods. More importantly, theories are understood as something that emerges from and with the objects of inquiry.

I do agree with Ahmed's critique that a certain 'founding gesture' can be found in some of the earlier work calling out 'a new materialism' or a 'material turn'. I do also agree with the problematization that some of this work might run the risk of erasing feminist genealogies by establishing a specific (counter) narrative that is only one story and how it could be told. With that being said, however, Ahmed herself not only fails to see that new materialisms, rather than "signifying a dismissal of feminism's legacies", could also be understood as "a reconfiguring of this inheritance and an opening up of new possibilities", as Noela Davis argues (2014: 63),[27] but also her own critique brings with it a number of problematic claims and misunderstandings. Not only that it is Ahmed who speaks of an alleged "feminist bio-phobia" where what, in fact, is criticized in much of new materialist literature are certain (feminist) social constructionist claims about biology and bodily materiality, Ahmed also makes the same mistake as some of the authors she criticizes, namely to construct a supposedly homogenous group "the new materialists" (Ahmed 2008: 32), she is then criticizing. But what weighs perhaps even heavier is that by claiming that "the new materialists" would construct "feminism as 'prohibiting'" (ibid: 24), Ahmed comes up with a highly problematic analogy between "these critiques of feminist anti-biologism" and "the evocation of political correctness as a form of prohibition against certain kinds of speech" (ibid: 31).

"In both these cases (biological arguments against anti-biologism, anti-pc arguments against pc), the speech act works by constructing an 'imaginary prohibition', which is then taken as foundational to a given speaking or intellectual community. That prohibition is imagined as hegemonic, as a majority position, in order to constitute the speech act, as a minority position, even as a kind of defence against free thinking, a rebellion against

---

27 Karen Barad puts forward a very similar idea, understanding new materialisms not as attempts at breaking with the past, but rather as "a dis/continuity, a cutting together-apart with a very rich history of feminist engagements with materialism" (Barad 2012a: 13). Feminist genealogy clearly matters here.

orthodoxy. The speech act that calls for us to 'return to biology' constructs the figure of the anti-biological feminist who won't allow us to engage with biology, and inflates her power." (Ahmed 2008: 31)

Not only that the point was never to "return to biology", neither in the sense of a return to biological determinism nor in the sense of a return to biology as discipline (leaving aside that there is no such thing as biology as discipline in singular), in the new materialist literature Ahmed is problematizing; in a certain way such an analogy also relativizes the racist strategy and practice of talking about 'political correctness' as prohibiting certain people to say what 'needs to be said', just to say it right away anyway. Ahmed is very well aware of the fact that "this is a potentially awful analogy" (Ahmed 2008: 31); and yet she sticks to it. What is more, this analogy rests on a misunderstanding of some of the central arguments of the works Ahmed is discussing. This problematic reading of new materialist encounters with social constructionist considerations of matter and material bodies becomes especially visible in Ahmed's understanding of the work of the feminist quantum physicist Karen Barad. Ahmed not only regards Barad as one of the key proponents of 'the new materialisms', even though, Barad uses the term 'new materialism' not once in her work,[28] but also accuses her of understanding Judith Butler as someone "who reduces matter to culture" (ibid: 32). Ironically, what Barad does is precisely not to read Butler in such a reductionist way but, on the contrary, to take up Butler's plea that what is needed would be a "return to the notion of matter, not as a site or surface, but as *a process of materialization that stabilizes over time to produce the effect of boundary, fixity, and surface we call matter*" (Butler 1993: 9). Thus, Barad draws productively on Butler's account of materialization, rethinking it with regard to the question of how not only discourse but also matter itself comes to matter. Ahmed's claim that Barad would turn matter into a "fetish object", "an it" and with that "reintroduces the binarism between materiality and culture" (ibid: 35), therefore, is simply wrong. What Barad is emphasizing over and over again is that in her "agential realist account, matter does not refer to a fixed substance;

---

28 In a personal conversation in February 2013, Karen Barad not only problematized the use of 'new materialism' as a label, but also the very idea of turns ('the new turn', 'the last turn', etc.) are in fact part of what she wants to overcome rather than endorse. Instead of turning away from something, Barad emphasizes the need for carefully (meaning, 'generously and respectfully') reading insights diffractively through one another in order to build theories, as particular ways of life, that have a chance for flourishing.

rather, matter is substance in its intra-active becoming—not a thing but a doing, a congealing of agency" (Barad 2007: 151).[29] Consequently, the accusation of Nikki Sullivan (2012: 309) who draws on Ahmed's paper, arguing that Barad would understand matter as "a priori" or as "matter *as such*", also fundamentally misses the point. Rather, matter and, for that matter, material bodies are conceptualized as *ongoing* intra-active material-discursive *processes* in Barad's account. Which is particularly interesting since, in the very same paper, with the notion of 'somatechnics', Sullivan offers not only a highly promising phenomenological approach for exploring the relationship between technologies (including technologies of the self) and the body, but also one that is not too far away from Barad's account.

Deriving from the ancient Greek words 'soma' (σῶμα) for body and 'techné' (τέχνη) for craftsmanship, the concept of somatechnics supplements the logic of the 'and', designating that technologies are not something 'we' add or apply to already given and constituted bodies. Rather, the concept indicates that soma (that is, "bodily-being-in-the-word") and techné (that is, "*dispositifs* in and through which corporealities, identities and differences are formed and transformed") are inextricably intertwined (Sullivan 2012: 302). Technologies are therefore always already enfleshed and corporealities are always already technologized. Somatization, thus, designates 'the coming-to-matter' of bodies for Sullivan. This idea, however, not only resonates with Donna Haraway's and Judith Butler's notion of materialization but, in fact, also with Barad's notion of materialization/mattering as well as her concept of the apparatus. In contrast to Barad, however, for Sullivan bodily materiality or "'matter' is inextricable from the I/eye that perceives it: perception makes 'matter' matter, it makes 'something' (that is no-thing) (un)become as such, it makes 'it' intelligible" (Sullivan 2012: 300).

In any case, the critique of a "return to science" or more specifically to biology (see Ahmed 2008; Bruining 2013) not only ignores that it were, precisely, feminist scholars trained in the natural sciences who have urged feminists and other critical scholars not to turn away from science and biology or to reduce them one-dimensionally to a kind of will to power,[30] but perhaps also tells more

---

29 I shall return to a detailed discussion of Karen Barad's agential realism in chapter three.

30 Lynda Birke (1999) goes even further here, arguing that biology is clearly "political" and precisely therefore has to be regarded as a "feminist issue". What is more, Birke even regards "biological knowledge" as a potential "feminist ally" (Åsberg/Birke 2010: 415). In *Reinventing Biology: Respect for Life and the Creation of Knowledge*

about the image of science that is prevalent in many social constructionist and poststructuralist approaches than about science itself. Dennis Bruining, for example, argues, expanding on Sara Ahmed's critique, that "new materialist considerations of matter often go hand in hand with a return to science as if contemporary scientific research has the answers postmodern(s) (feminists) have obscured with their obsessive focus on language and representation" (Bruining 2013: 162). In doing so however, Bruining not only seems to reproduce the idea that scientific theories and practices would only represent mere manifestations of instrumental reason, rationalizing and legitimizing dominance and social inequalities, but also seems to ignore that many new materialist feminist scholars are both 'postmodern' philosophers *and* trained in the sciences. Donna Haraway, who is trained in molecular and developmental biology, or Karen Barad who has a training in theoretical particle physics, *do* consider themselves to be profoundly shaped by their training as natural scientists. Haraway, for example, writes that she is,

"in love with biology—the discourse and the beings, the way of knowing and the world known through those practices. Biology is relentlessly historical all the way down. There is no border where evolution ends and history begins, where genes stop and environment takes up, where culture rules and nature submits, or vice versa. Instead, there are turtles upon turtles of naturecultures all the way down." (Haraway 2004: 2)

In a recent interview, Haraway not only speaks of herself as someone who "spent a lot of time in labs, with other organisms" but even problematizes that the commerce between (feminist) social constructionist and postmodern theories on the one side and the natural sciences on the other side "was never quite symmetrical [...] I learned a whole lot more about semiotics and psychoanalysis than folks learned about biology" (Haraway/Williams 2009). Karen Barad puts forward a similar argument, saying that,

"[i]t is no coincidence that so many feminist science studies scholars have been trained as scientists and that we have not shied away from expressing our deep love for science and this astonishingly remarkable, intricate, amazing world of which we are a part. Photons and electrons leave me in awe. So too brittlestars, humans, ferns, and so much more,

---

(1995), Lynda Birke and Ruth Hubbard point out how not only the practices of biology but also the very knowledge biological theories and practices produce would fundamentally change as soon as the objects of research, that is, animals, or more generally, the organisms concerned are considered as subjects with agency.

always more. These critters do more to deconstruct science (and others of our assumptions) than anything going under the name of social constructivism." (Barad 2011: 450)

Bruining does state that "the engagement with science itself [...] *can* certainly be useful" (ibid: 158, italics JB). But again, science seems to be something that *can* only be 'useful' as the object of critique for 'postmoderns' and 'feminists', rather than the source for critical theories. Science however equals mere ideology just as little as biology is identical with biological determinism and fatalism. Not only that 'biology' can mean very different things, such an equation can also only work if 'biology' is pictured as opposed to culture, as static and fixed. Donna Haraway reminds us that denoting processes of being, biology is "a complex web of semiotic-material practices" (Haraway 1997: 217); while as a discipline, biology is "knowledge producing practice" (ibid: 218). For someone who has dealt a good deal with the history and the methods of the life sciences, it is frustrating to see that the picture of the natural sciences as a homogenous apparatus of surveillance and oppression still lies at the heart of some feminist and other critical theories, implying that critical thought and practice can only be pursued 'outside' the sciences – even though feminist scholars trained as scientists such as Donna Haraway, Evelyn Fox-Keller, Anne Fausto-Sterling, Lynda Birke, Karan Barad, Ruth Hubbard, and many others, have demonstrated that this story could also be told differently.

Taking into account questions of material or bodily agency is not the same as falling back into biological determinism, just as problematizing certain claims of social constructionist understandings of technology, such as, for example, the idea that technology can be reduced to or even equated with social (power) relations, does not mean to bring back technological determinism. Therefore, it is not about a supposed 're/turn to science' or about declaring society and the social irrelevant but, on the contrary, about the question of how to consider materiality and material or bodily agency beyond a reductionist perspective that understands bodies either as brute matter or primarily as effects of powerful discourses and technologies (including technologies of the self).[31] Is there a way to take into

---

31 As a matter of fact, Foucault does not differentiate sharply between material technology and technology as practice. In one of his last interviews, Foucault problematizes the "very narrow" meaning of the word 'technology': "A very narrow meaning is given to 'technology': one thinks of hard technology, the technology of wood, of fire, of electricity. Whereas government is also a function of technology: the government of individuals, the government of souls, the government of the self by the self, the government of families, the government of children and so on" (Foucault 1984: 256).

account how the material and the discursive, the natural and the cultural, the biological and the social, as well as bodies and technologies fold into each other, where it is not clear what produces what; rather than already departing from the idea of preexisting entities that precede their entanglements, interacting with one another?

There is no doubt that the body has been, and certainly continues to be, central for feminist and queer scholars, scholars of color, and dis/abled scholars, to name but a few. The question however remains which body?[32] As illustrated above, social and linguistic constructionist theories do not deny the existence of a material body. Such an accusation would indeed be a caricature of a rich and long constructionist debate on the body. Rather, the critique was always that in much of social constructionist literature the body first and foremost appears as a mere inert object or even worse is reduced to its surface and with that is regarded as a blank slate for social and cultural inscriptions. Consequently, the body behind or beyond (not before) its socio-cultural and discursive representation all too often remains hidden. Therefore the body is present and yet absent, everywhere and nowhere at the same time. Matter, as something that is constantly in motion, constantly in becoming, disappears behind the endless proliferation of discourses. This fallacy might not only be the price for social constructionisms necessary anti-essentialism but also the legacy of its transcendental idealist roots.[33] The belief that matter can neither organize itself nor does it have any kind of "spontaneity" and activity, lies at the heart of Kant's epistemology. For Kant "the possibility of living matter cannot even be thought", for its very concept "involves a contradiction, because lifelessness, inertia, constitutes the essential character of matter" (Kant 1951 [1790]: 242; § 73). It is precisely this belief that matter and therefore also material bodies are inert, mute objects awaiting the

---

Hence, in a certain sense it could be said that for Foucault technology seems to be more of a practice, a doing, that is, a process, rather than a mere artifact or system.

32 Terry Eagleton puts it in a similar, albeit perhaps also too generalized way, saying that "[p]ostmodernism is obsessed by the body and terrified of biology. The body is a wildly popular topic in U.S. cultural studies – but this is the plastic, remouldable, socially constructed body, not the piece of matter that sickens and dies." (Eagleton 2003: 186)

33 Ian Hacking, although polemically, even traces back the roots of social constructionism to Plato: "Surprise, Surprise! All construct-isms dwell in the dichotomy between appearance and reality set up by Plato and given a definitive form by Kant. Although social constructionists bask in the sun they call post-modernism, they are really very old-fashioned." (Hacking 1999: 49)

power of language and the mind in order to be filled up with life that social constructionism still carries with it.

As a result, social constructionism bears the risk of confusing social processes of construction with the material reality itself.[34] Social constructionism "confuse[s] what the world is made of with how it is made [...] the ingredients with the construction. They believe that the world is really made of social stuff" (Latour 2003: 26). The idea that material objects such as, for example, certain technologies and even bodies would be "constructed out of worldviews", as Sergio Sismondo (1993: 566) problematizes, not only seems a too reductionist perspective but also leads to the contradiction I have outlined above, namely, that while technology, the body, and even material reality itself, is considered as socially constructed there still remains a non-social nature or material realm behind or beyond its construction. This reality behind or beyond its social construction, however, remains an essentialistic, ahistorical, passive, and static one. Thus, although at first glance the idealist mind-set lying at the roots of social constructionism liquefies materiality and facticity, social constructionism is far from breaking with classical metaphysics. In fact, it implicitly reinstalls an essentialist ontology through the backdoor for everything that lies 'outside' the social realm. Instead of getting rid of the idea of a nature 'out there' it is only turned into a nature behind or beyond its social and cultural constructions, a nature that is buried deep under the endless proliferation of discourses and the processes of social construction, but still remains intact as antithesis to culture and the mind. It is also for this reason that social constructionism is, in fact, a metaphysical position, because it is "directed at certain pictures of reality, truth, discovery, and necessity. It joins hands very naturally with what Nelson Goodman calls irrealism: not realism, not anti-realism, but an indifference to such questions, which in itself is a metaphysical stance." (Hacking 1999: 60-61)

"Social constructionism got foregrounded", Donna Haraway states in a recent interview, "the volume got turned way up and it got developed in all kind of creative ways. But then it becomes sclerotic. It becomes an orthodoxy" (Haraway 2005: 155). What if, it is true that what is happening to social constructionism today is very similar to what happened to positivism in the middle

---

34 This does not mean to assume that material reality and social or cultural reality would represent two distinct realms. On the contrary, it seems considerably more promising to understand the natural and the social, the material and the discursive as always already inseparably entangled with one another and as not preceding their 'meetings', as I will make explicit in the following chapters.

of the 20th century, transforming it from a once critical tool into orthodoxy,[35] blocking the possibility for new concepts, theories, and perspectives that might lead to a more liveable future for everyone? Social constructionism cannot be and, certainly, is not the end or the final stage of critical thought. But precisely because (feminist) social constructionist theories were highly successful in critiquing naïve realism and essentialism, especially, when it came to questions of social inequality and oppression, the goal cannot just be to discard social constructionism all together. Rather, my aim in what follows lies in transcending and yet incorporating (feminist) social constructionist insights, in order to be able 'to get elsewhere'.[36]

The body might not exist apart from its social and cultural construction, yet it is not reducible to social and discursive practices and relations alone. Rather than being inert matter the body as a relational entity has its own forces. Therefore, there is a need for theories that allow us to take into account the non-discursive (not pre-discursive) facticity of bodies in their material-discursive movements.

---

35 Nikki Sullivan's statement toward new materialist scholars drawing on the work of Bruno Latour for critiquing certain claims of social constructionist though may be regarded in this light. Sullivan argues that although it "would perhaps make sense at this point in the article to turn to the work of Latour and engage directly with the concepts cited above", for someone "whose ways of seeing, knowing, being, feeling and so on have been profoundly shaped by feminism and poststructuralism I am both sceptical about, and uncomfortable with, the idea that we must turn to the voice of authority (in the guise of science or male philosophers) in order to see clearly, and so for strategic ethico-political reasons, I will resist such a move" (Sullivan 2012: 307-308). I do share the concern about the pressure of having to refer to the work of (generally male) authorities in the sciences and humanities. That being said, however, what Sullivan does here is not only to implicitly deny critics of social constructionism to be feminist and poststructuralist too but she also implies by creating a straw man ("the voice of authority") that there is no need to engage with the critique of certain assumptions (in this case those of social constructionist and poststructuralist theories) as long as one identifies oneself with the theories and positions under critique and is "profoundly shaped" by them. By immunizing the own position against critique and making it non-negotiable, such a position comes critically near to orthodoxy.

36 In an interview, Donna Haraway makes clear that she refuses both naturalism and social constructionism; rather, the aim has to be to get elsewhere. "I am neither a naturalist nor a social constructionist. Neither-nor. This is not social constructionism, and it is not technoscientific, or biological determinism. It is not culture. It is truly about a serious historical effort to get elsewhere." (Haraway 2003a: 55)

Understanding the body as agentic might allow us to come to different understandings of how and why technologies (again, including technologies of the self) can fail in their attempts at reconfiguring material bodies for the demands of the political economy – precisely, because "the body itself is anything but a blank page for social inscriptions, including those of biological discourse", as Donna Haraway (1991a: 197) urges us to understand. The aim that lies at the heart of the following chapters, hence, will be to find a way to draw productively on this idea of the body and at the same time taking it a step further in order to be able not only to come to an understanding of bodies as always already material-discursive assemblages of natural-cultural-technological forces, but also to shift the focus even more onto the very processes through which bodies come to matter, in both senses of the word – that is, not only as meaning but also as fluid matter. Certainly, such an account has to start by acknowledging that there is no such thing as *the* body but only bodies in the plural that are not only always marked by race, sex, gender, class, ethnicity, nationality, dis/ability, species, and much more, but also constantly in becoming.

But how to grasp material bodies as agents when technologies and technoscientific practices come into play without losing sight of important questions of power and inequality? How can the body no longer be conceived as plastic matter upon which technologies act but instead as something that resembles much of what Michel Serres (1982) has termed a thermal exciter, an entity that is "both the atom of a relation and the production of a change in this relation",[37] and hence as potentially unruly and actively involved in the process of its own becoming? I believe that shifting the focus onto the question of how bodies come to matter and what bodies can do rather than what the supposedly natural body is, allows for such an understanding of bodies as endowed with the capacity for transformation, and hence for an understanding that acknowledges that "the 'flu-

---

37 See the translator's introduction in *The Parasite* (Serres 1982: x). Later in the book, Serres uses the term (thermal) exciter in order to describe what a parasite can do. "The parasite is an exciter. Far from transforming a system, changing its nature, its form, its elements, its relations and its pathways […], the parasite makes it change states differently. It inclines it. It makes the equilibrium of the energetic distribution fluctuate. It dopes it. It irritates it. It inflames it. Often this inclination has no effect. But it can produce gigantic ones by chain reactions or reproduction. […] The parasite intervenes, enters the system as an element of fluctuation. It excites it or incites it; it puts it into motion, or it paralyzes it. It changes its state, changes its energetic state, its displacements and condensations." (Serres 1982: 191)

idities and fracturings' of the body may offer the promise of endless transformation, thereby making it attractive to feminist theorists" (Davis 2007: 55-56).

2 Locating the Technological with/in Rhizomatic Networks

> There is nothing very mysterious about how to identify important philosophical problems concerning technology. They press on us in such a way that avoiding them seems the more puzzling problem.
> —Donna Haraway/*Philosophy of Technology: 5 Questions*

If Derrida is right with the claim that technology is not the Other of the body, the question what then technology is gains center stage. This question becomes even more complicated, if technology is not reduced to its mere material aspects but also understood as a particular mode of knowing and being in the world as part of it. In *Pandora's Hope*, Bruno Latour writes that technology does not exist as such, "that there is nothing that we can define philosophically or sociologically as an object, as an artifact or a piece of technology" (Latour 1999a: 190-191).[1] In doing so, Latour even goes so far as to argue that "a nuclear plant, a drone [...] are no more made *of* technology than law is made *of* law or religion *of* religion" (Latour 2013: 212), precisely because technology seems to have no material

---

1   In his more recent work, Latour reinforces this idea, saying that "[w]e need to see 'TECHNIQUE' and 'TECHNOLOGY' not in their noun forms but as adjectives ('that's a technical issue'), adverbs ('that's technically/technologically feasible'), even sometimes, though less often, in verb form ('to technologize'). In other words, 'technology' does not designate an object but rather a difference, an entirely new exploration of being-as-other, a new declension of alterity. Simondon, too, made fun of substantionalism, which, here again, here as always, failed to grasp the technological being" (Latour 2013: 223).

essence. Instead of understanding technology as always already what we mean when we talk about technology,[2] Latour suggests to speak of technology whenever we are witnessing processes of mutual mobilizations of human beings and nonhuman entities through which properties get exchanged in both directions. This does not, however, mean that humans would be increasingly objectified and technical objects socialized, but rather that both humans and technical objects are coming ever closer, with the effect that the categorical distinction between the two can no longer be upheld as easily as it is often thought. Actions, politics, and even moral beliefs are delegated to things through the exchange of properties, altering their form and function. But also the *programs of action* delegated to things themselves change before they, in return, influence human actions and thoughts. Influenced by Michel Serres for whom the classical model of communication as a linear transmission of information seemed to be a most unlikely case, Latour understands transmissions always as transformations and reconfigurations. This is precisely the backdrop against which Latour speaks of technical mediation, expressing that technologies constitute mediators "endowed with the capacity to translate what they transport, to redefine it, redeploy it, and also to betray it" (Latour 1993: 81). A mediator, for Latour, is something that interrupts, alters, causes complications, and transforms; it is both a means and an end. However, it is important to understand that mediation is not simply something that happens between two elements. Rather, it is the emergence of something new out of something existing; changing both, the new and the old.[3] Does Latour's concept of technical mediation provide us not only with an account that understands technology as a process rather than as a substance or an already given entity, but also help us to contest prevalent understandings of technology as an expression or a manifestation of technological rationality, and hence as

---

2   In a similar way, Trevor Pinch et al. (1992: 285, Fn. 4) argue that it could be said that technologies become technologies through a specific rhetoric. 'Rhetoric', here, means "a set of systematically used recurring textual features whereby texts gain their persuasive power".

3   For this reason, as it will become clearer in the next chapters, why it also makes little sense to speak of technical mediation as a process through which something previously 'natural' and supposedly unmediated now would become *technologically* mediated and hence fundamentally altered in its ontological state and meaning. Technical mediation is no one-way-street. See, for example, Barbara Duden's *Disembodying Women* (1993) for such a one-dimensional understanding of technical mediation using the example of medical technologies, and how they would fundamentally alter a supposedly natural body and bodily functions.

something that is opposed to the natural body – yet without understanding technology ahistorically and as being neutral but, on the contrary, as always lying at the heart of processes of material reconfigurings?

THE MODE OF EXISTENCE OF TECHNICAL OBJECTS

Latour's notion of the mediator can be traced back to Gilbert Simondon's philosophy of technology. Already in the 1950s, Simondon contested both the hegemonic philosophical understanding of technology as an extra-social power, determining the fate of society, and the critique of technology as a mere expression or manifestation of instrumental reason. Even though Simondon's considerations have been highly influential for a number of philosophers such as Michel Foucault, Gilles Deleuze and Félix Guattari, Michel Serres Bruno Latour, Isabell Stengers, and Andrew Feenberg,[4] his work remained relatively unnoticed for a long time.[5] In contrast to many of his contemporaries such as Lewis Mumford, Martin Heidegger, and Herbert Marcuse, Simondon was one of the very few philosophers at his time who has critically engaged with the *Macy Cybernetics Conferences* which took place in New York City between the years 1946 and 1953. Sponsored by the Josiah Macy, Jr. Foundation, the conferences aimed at promoting interdisciplinary exchange between the natural sciences, the humanities, and the social sciences. Among the participants were prominent scholars such as Norbert Wiener, Heinz von Foerster, Gregory Bateson, Oskar Morgenstern, and Paul Lazarsfeld. As one of the most important scientific events after the Second World War, the Macy Conferences on cybernetics can be regarded as marking the epoch-making step from thermodynamics to cybernetics, from disciplinary societies to societies of control, and from the industrial society to the information society.[6]

---

4 In an interview, Andrew Feenberg (2004) states that "[t]he most neglected important figure is Gilbert Simondon, whose work has influenced me".

5 It was not before the year 2012 that Simondon's main work *Du mode d'existence des objets techniques* from 1958 was translated for the first time into German and four years later into English. Simondon's philosophy has been rediscovered in the last few years and is now increasingly taken up by philosophers of technology and (feminist) science studies scholars. See, for example, the work of Pascal Chabot (2013); Muriel Combes (2013); Arne de Boever et al. (2013); Henning Schmidgen (2012); as well as Iris van der Tuin and Aud Sissel Hoel (2012).

6 For a detailed study on the Macy Cybernetics Conferences, see Pias (2004).

The term cybernetics (κυβερναν, meaning the art of steersmanship) was coined by the mathematician Norbert Wiener in "Cybernetics: Or Control and Communication in the Animal and the Machine" (1948), referring to self-regulating mechanism in organisms and machines. Emerging out of military research during the Second World War, cybernetics focused on questions of regulation, control, and information processing in technical and organic systems. Wiener himself was involved in the development of automated fire control systems for anti-aircraft guns to prevent German air strikes against allied cities and other targets. Less known, however, is Wiener's commitment against the military-industrial complex after the Second World War. In 1946, already a year before his letter to the *Atlantic Monthly* entitled "A Scientist Rebels" stirred up the dust, Wiener wrote a number of brief letters addressed to scientists and engineers who were working for the military-industrial complex and who had asked him for scientific advise. In a letter to a Mr. Forsythe from the Physical Research Unit of Boeing Aircraft Company, Wiener writes,

"Since the termination of the war I have highly regretted the large percentage of scientific effort in this country which is being put into the preparation of the next calamity. I therefore am much gratified to find that my publication on 'Extrapolation, Interpolation, and Filtering of Stationary Time Series' is no longer available to those who construct controlled missiles. I can, of course, furnish you with no advice as to where to find them." (Wiener/Pach 1983 [1946]: 36)

In a subsequent letter Wiener makes his position even clearer, emphasizing that the bombing of Hiroshima and Nagasaki has made obvious that being a scientist and providing scientific information to the military "is not a necessarily innocent act, and may entail the gravest consequences. [...] The experience of the scientists who have worked on the atomic bomb has indicated that in any investigation of this kind the scientist ends with the responsibility for having put unlimited powers in the hands of the people whom he is least inclined to trust with their use" (Wiener/Pach 1983 [1946]: 37). For Wiener it was clear that weapons, once they have been developed, "will be used" eventually. It has to be understood in this light why Wiener protested against the militaristic use of his work on guided missiles, knowing that he still could not prevent its application for the development of new weapons.

"The practical use of guided missiles can only be to kill foreign civilians indiscriminately, and it furnishes no protection whatsoever to civilians in this country [...] If therefore I do not desire to participate in the bombing or poisoning of defenseless peoples—and I most certainly do not—I must take a serious responsibility as to those to whom I disclose my scientific ideas. Since it is obvious that with sufficient effort you can obtain my material, even though it is out of print, I can only protest 'pro forma' in refusing to give you any information concerning my past work [...] I do not expect to publish any future work of mine which may do damage in the hands of irresponsible militarists." (Wiener/Pach 1983: 37)

What is evident here is not only Wiener's refusal to take part in military research – Wiener indeed refused any military funding for his work after the Second World War[7] – but also his plea toward other scientists not to take part in military research that promotes and rationalizes the (automated) killing of people.[8]

In his later work, Wiener shifts his attention even more to epistemological and political questions, bringing to the fore the social implications of cybernetics. Already in the early 1950s, Wiener was convinced that in the very near future computers would be not only used for scientific calculations but would also eventually revolutionize the whole production process, leading to a new industrial revolution, or perhaps more precisely a revolution in information and communication. Wiener saw this new revolution as fundamentally characterized by automated machines which would act "as a source of control and a source of communication. We communicate with the machine and the machine communicates with us. Machines communicate with one another. Energy and power are not the proper concepts to describe this new phenomenon" (Wiener 2003 [1954]: 71). Arguing that many engineers would never think "further than the construction of the gadget", leaving open "the question of the integration between the gadget and human beings in society" (ibid), Wiener argues that there is a necessity to understand the machine and how it works as something that is not external to the human.

"If we want to live with the machine, we must understand the machine, we must not worship the machine. We must make a great many changes in the way we live with other people. We must value leisure. We must turn the great leaders of business, of industry, of politics, into a state of mind in which they will consider the leisure of people as their business

---

7   See the Introduction to "Men, Machines, and the World About" in Wiener (2003 [1954]: 65).
8   In "A Scientist Rebels", Wiener explicitly says that the aim of his letter is also to get the attention of other scientists in the hope that they "make their own independent decisions, if similar situations should confront them" (Wiener 1947: 31).

and not as something to be passed off as none of their business." (Wiener 2003 [1954]: 71.)

Beside Wiener's emphasis on the value leisure, what becomes apparent here is a rethinking of the relation between the organic and the mechanical, between human beings and machines. Rather than being a mere tool to be feared or worshiped the machine becomes something that is always already entangled with human beings in different ways.

It is precisely this idea Gilbert Simondon follows up in his philosophy. Incorporating but also problematizing the insights of early cybernetics, Simondon develops in *Du mode d'existence des objets techniques* (1958) a philosophy of technical objects[9] that tears down the opposition between the cultural and the technical as well as between human beings and machines. For Simondon, early cybernetics blurs the line between organisms and machines as well as between the natural sciences and the humanities, and in doing so provides a new, promising account for philosophical studies of technical objects. On the other hand, however, by understanding environmental influences as disturbances of an otherwise hermetically sealed off system, cybernetics would ultimately fail to break with the belief that the natural and the artificial, and thus technical objects and human beings, represent two different ontological spheres.

Against this backdrop, Simondon emphasizes that the industrialized societies of the early 20th century had entered into a new regime of mechanism, in a new "réalité gouvernée" (Simondon 1958: 14) that covers both human beings and technical objects alike. Although this new regime can be characterized as one in which technology significantly influences and regulates the social, to an extent that it can be said that today's societies are thoroughly technological, an inability to reflexively process this shift prevails. Instead, a pre-industrial view on technology has been preserved which manifests in the asymmetry between modern technology on the one side and its socio-cultural assessment on the other. For Simondon, the main characteristic of this pre-industrial view on technology can be found in the idea of technical objects as mere tools or means to an end, and therefore as something detached from human beings, as something foreign, as the Other.

---

9   It is important to understand that even though Simondon speaks of technical *objects* he does not understand the term object opposed to the term subject. Rather object and subject are thought of as different moments of the same phenomenon since there is no *substantial* difference between the natural/organic and the artificial/mechanical for Simondon (see Combes 2013; especially the chapter on transindividual relation).

According to Simondon, the understanding of technical objects as something foreign reveals itself in the juxtaposition of technology and society, and culminates in the idea of a dissolution of nature and an alienation of human beings in the wake of the machine that pervades conservative and critical philosophies of technology alike. It is in this light that for both Heidegger and for Adorno, modern technology largely represents a mere instrument of domination over Nature. While for Heidegger (1977) modern technology reveals (*entbirgt*) and challenges or sets upon (*stellt*) nature, and in doing so degrades it to a mere resource, Adorno (2003 [1953]: 316-317; trans. JB) considers technology as "not the primary societal nature, not the matter itself, not humanity, but only something derived, the organizational form of human work" which aims at the domination of 'inner' and 'outer' nature.[10] Simondon, on the other hand, argues that it would be a mistake to attribute the alienation of workers to the machine itself or to certain qualities of technology, rather it has to be understood as the effect of the degradation of workers to mere operators of the machine. In doing so, Simondon reverses the idea that technical objects would determine and dominate human beings. As a matter of fact, for Simondon, it is the machine that is treated like a slave. The reason for this would lie in the common belief that technical objects contain no "human reality" (Simondon 1958: 9). The belief that there is no humanity in technologies, that technologies contain no "human reality", would then be what leads to the idea that the categories of the humane and the natural must be defended against the invasive and everything-pervading artificial and mechanical at all costs. What follows from this, for Simondon, is not only the call to philosophically take into account technical objects but also a politically and morally irritating analogy that he draws between the fight against human slavery and the necessity of freeing technical objects from their instrumentalization. Departing from this idea, Simondon develops a philosophy of technology that does not separate technology from the human. It is important to understand however that Simondon does not intend to reduce human beings on technical objects or, conversely, to grant technical objects human qualities. In fact, influenced by the vitalist philosophy of Georges Canguilhelm and Norbert Wiener's thoughts on cybernetics, Simondon rather raises the question for the ontological state of technology by turning to the machine – this time, however, without understanding technology a priori as something external to the human (body). While

---

10 Just as for Simondon, it is also clear for Latour that such a perspective denies being as essentially 'technological' (Latour 2013: 220) even *before* any normative political and moral assumptions are made with respect to that very being, and the question how its 'nature' ought to be conceptualized.

Heidegger (1977: 4) argued that "the essence of technology is by no means anything technological" and subsequently raised the question '*what* technology then is', Simondon shifts the focus to the question *how* technologies function, how they relate to other technologies and human beings.

Technical objects undergo a development from the abstract to the concrete adopting more and more functions, progressively condensing more potentials.[11] The development of technical objects consequently marks a mutation, however one that has a "meaningful direction" (*mutation orientée*) (ibid: 40). For Simondon, even though machines and tools are made and controlled by human beings they also have a life of their own, a specific mode of existence. Technological development, thus, has to be understood as a quasi-autonomous project. In contrast to technological determinist theories, Simondon unfolds a philosophy within which technical objects are always part of a heterogeneous orchestra conducted by human beings. As such, technologies can never fully determine our lives and bodies. What follows from this thought is the programmatic assumption that the machine is "a stranger to use [...] a stranger in which what is human is locked in, unrecognized, materialized, and enslaved, but nonetheless remains human" (Simondon 1958: 9; trans. JB). As for Simondon human history, human ideas and practices, always form a part of the machine, 'the human' persists in technical objects. What is inherent to technical objects, thus, is nothing less than a "human reality" which resides in machines as human actions "fixed and crystalized in functioning structures" (ibid: 12). Half a century later Bruno Latour will take up this idea when he emphasizes that "[p]rotecting human beings from the domination of machines and technocrats is a laudable enterprise, but if the machines are full of human beings who find their salvation there, such a protection is mere absurd" (Latour 1993: 164).

Despite the fact that technical objects do have a certain kind of agency and thus a mode of existence which differs from that of human beings, for Simondon, they nevertheless need the human as a "permanent organizer" and "living inventor". But since technical objects are also always radically open, that is, undetermined, it is never clear in advance what they can achieve and in which

---

11 Andrew Feenberg takes up this thought, adding to Simondon's notion of "technological concretization" the idea of "social concretization" in order to be able to take into account "socialist demands for environmentally sound technology and humane, democratic, and safe work", and, in doing so, to illustrate that they "are not extrinsic to the logic of technology but respond to the inner tendency of technical development to construct synergistic totalities of natural, human and technical elements" (Feenberg 2010: 192).

direction they might develop. What is more, this development is never one that could be separated from the development of society; and still it is neither society that adopts technologies as if the latter would represent mere means to an end, nor is it technology that determines society as a kind of autonomous force. In fact, it is exactly this openness and open-endedness through which technical objects and human beings are overlapping and folding into each other. As mentioned earlier, it is important to understand that through these processes neither technical objects are humanized nor are human beings objectified; rather, human beings and technical objects constitute a fabric which Simondon designates with the term "technical ensemble". Humans 'move' the machines but at the same time are also always themself 'moved' by the machines. As part of the technical ensemble the machine appears as the opposite of an alienating force. In its function to oppose and slow down entropy the machine has to be seen as an entity that stabilizes our world temporarily by bringing order into it. Simondon makes clear that he refuses to regard the artificial and the organic as interchangeable with regard to their modes of functioning, as, for example, cybernetics did.[12] Human beings and machines have to be understood as ontologically different but not as substantially different since both human beings and technical objects, the organic and the mechanical, would share the same effort to increase negentropy.

"La machine, comme élément de l'ensemble technique, devient ce qui augmente la quantité d'information, ce qui accroît la négentropie, ce qui s'oppose à la dégradation de l'énergie, […] ce qui s'oppose au désordre, au nivellement de toutes choses tendant à priver l'univers de pouvoirs de changement. La machine est ce par quoi l'homme s'oppose à la mort de l'univers." (Simondon 1958: 15-16)

The machine as an element of the technical ensemble is what augments the quantity of information, what increases negentropy, what opposes the degradation of energy, Simondon writes. Thus, it is the machine that "fights against the death of the universe" by slowing it down. With the term "negentropy" or negative entropy, Simondon, draws on a concept of Erwin Schrödinger (2006 [1944]: 73) who applied it for the faculty of a living organism to "feed upon negative entropy", that is, to 'consume' order from its environment in order to be able to

---

12 Referring to the philosophy of Henri Bergson, Norbert Wiener, for example, argues that, "the modern automaton exists in the same sort of Bergsonian time as the living organism; and hence there is no reason in Bergson's considerations why the essential mode of functioning of the living organism should not be the same as that of the automaton of the same type" (Wiener 1948: 44).

maintain "itself stationary at a fairly high level of orderliness (= fairly low level of entropy)".[13]

Despite the fact that Simondon takes into account a number of concrete technical objects, such as electronic tubes and combustion engines, dismantling them and drawing highly accurate drawings to the extent that he is often referred to as that philosopher of technology who certainly cannot be criticized for *technologically* not knowing what he is speaking of (see, for example, Barthélémy 2011), his philosophical considerations on technology, precisely because of their generalizing character, seem to be less helpful for political questions concerning technologies and bodies in their entanglement with one another. After all, from such a perspective it could be argued that guided missiles, drones, nuclear bombs, or, for example, the bulldozers logging the rainforests at this very moment would not be fundamentally different in their attempt at decreasing entropy and thus producing 'order'. To be fair, Simondon's understanding of 'order' is a far more complex one, and he also does state that it would be a mistake to understand technology exclusively as a positive and productive force, for it also represents a will for power that is expressed in "the technicist and technocratic excessiveness of the thermodynamic age, which has taken a direction both prophetic and cataclysmal" (Simondon 1958: 15; trans. JB). And yet, in the end, Simondon's understanding of technology remains perhaps too generalizing to be helpful to understand technologies not only in their entanglements with material bodies but also in their ambivalences with regard to their ethical and political consequences.

---

13 In his recent book, *The Neganthropocene*, Berard Stiegler (2018) takes up Simondon's considerations on technology and Schrödinger's understanding of negentropy, reframing them against the backdrop of both the 'Age of Man', that is, the Anthropocene, and the age of digital media, as "neganthropy". "Organic life is that which defers that entropy described by Clausius on the basis of the works of Sadi Carnot, and in relation to which Schrödinger showed that every form of life is the local formation of a counter-tendency, which he called negative entropy. Exosomatization is the continuation of this process, but in a new sense, producing an increase of entropy and of what results from it, disorder, but also a new form of negentropy, which I call 'neganthropy', that is, the production of those new forms of locality that are, precisely, exorganisms. Tomorrow's challenge is to increase neganthropy, and to develop an economy that valorizes it systemically." (Stiegler 2018: 127-128) Thinking 'exosomatization' in terms of 'neganthropology', Stiegler not only actualizes Schrödinger, but also implies that the so-called digital age opens up a perspective of a different economy, one based on negentropy as a new value that might overcome the connection between use value and exchange value.

TECHNICAL MEDIATION AS PROCESSES OF MUTUAL MOBILIZATION

Just as for Simondon, who argues that the human resides *midst among* the machines that operate *with* him, it is also clear for Latour that "without technological detour, the properly human cannot exist" (Latour 2002: 252) Even more than for Simondon, however, Latour emphasizes that the philosophy of technology has to shift its view to concrete technologies in their complex entanglements with other living and non-living forces and entities. What is more, for Latour it is never clear in advance what happens with/in these entanglements. In and through technologies political and moral programs of action – defined by Latour (1999a: 178) as "series of goals and steps and intentions" – manifest, producing certain effects. Latour elaborates this idea by drawing on several examples; amongst many others is the Berlin key (Latour 1991b).

At the beginning of the 20th century, Berlin, like many other large European cities, had to deal with a high rate of delinquency due to poverty and an accelerated migration from the countryside. Thus, the residents of tenement houses, often dismissively called 'rental barracks',[14] had been asked to lock the entrance doors leading into the inner courtyard in the evening in order to avoid uninvited guests. However, since many residents would not follow these instructions, a locksmith had to come up with a material solution for this problem: the Berlin key. Once the entrance doors had been locked by the caretaker in the evening,

---

14 Even though the use of the term 'rental barracks' instead of the more common term 'tenement houses' sounds rather militaristic, in "The Rental Barracks", a radio broadcast from the year 1930, Walter Benjamin emphasizes how close the rise of the tenement houses in Berlin was in fact tied to the sphere of the military. Berlin has been a military city for several centuries. Even under the reign of Frederick William I. "every Berliner was obliged to put up a certain number of soldiers, depending on the size of his house or apartment" (Benjamin 2014: 57). However, by the turn to the 20th century it proved to be increasingly difficult to accommodate all the soldiers and their families. At times the soldiers and their families constituted up to one third of Berlins population. As Benjamin explains, the reason for letting the soldiers live together with their families could be found in the fact that many soldiers deserted because of the inhumane conditions in the Prussian military. Letting them live together with their families – even if only in specific buildings that they were not allowed to leave – was thought of as a way to lower the chance for deserting. This was the birth of the massive buildings called 'rental barracks'. Entire streets had been packed with giant buildings erecting into the sky and only occasionally intersected by streets and open space.

tenants, if they wanted to enter or leave the house had not only to unlock the door but also to lock it in the very same action from the other side in order to get their keys back. This was necessary because, in contrast to ordinary keys, the Berlin key had two key blades, one at each end of the key. Meaning that once the key had been inserted in the keyhole to unlock the door the key owner was forced to push the key all the way through the lock in order to be able to recover it on the other side, only after turning the key again for 270 degrees counter-clockwise and with that locking the door. The imperative to lock the door at night had been molded into the form and function of the key and the lock. With it, also political and moral imperatives had been inscribed in the key, namely to lock the door at night to prevent uninvited guests from the house. Since oral and written instructions to keep the doors leading into the courtyards locked at night were not powerful enough to make the tenants follow them, they had been cast into iron. With that action, however, the very matter of expression had been changed from words to things, and the Berlin key was born. What Latour makes explicit is nothing less than that the social cannot be built from the social alone; it always needs things through which it obtains its stability and durability. Taking only the social into account by focusing primarily on humans in their actions and interactions misses the fact that technical objects too have morals in the sense that they embody specific social norms, interests, and rules.

In contrast to the social constructionist idea of an inscription of politics or social norms and interests into technologies, there is no preexisting 'social' for Latour that could be inscribed into technologies in a second step (for example, during decisions concerning the design and construction of technologies) with the consequence that technologies would ultimately represent mere placeholders for 'the social' or congealed social actions and interests. Where social constructionist theories of technology emphasize that there is no such thing as an internal technical logic that would drive technical objects in certain directions but rather that it is 'the social' that shapes technology (see chapter one), for Latour, 'the social' itself is something that has to be explained in the first place. What is more, there is no guarantee that the programs of action, materializing temporarily in and through technologies, could achieve the effects anticipated by particular social actors. Programs of action are indeed delegated to nonhuman actors, but these actors do also have the ability to bypass the programs delegated to them and even to establish anti-programs. Therefore, it is never clear in advance what happens to the delegated programs of action after their inscription into technical objects. Precisely because if it were that easy it would be enough to develop technologies that could determine and discipline us in our everyday lives to such an extent that we would have no other choice than to painstakingly follow the programs of actions inscribed into them. Thus, programs of action must not be understood deterministically. Because what, in fact, happens to them after they are delegated to nonhuman entities always depends on the manifold entanglements through which a multitude of actors are folded into one another. To put it in a nutshell, there are never only keys, locks, and their disciplined owners.

Against this backdrop, analyses of technology focusing primarily on questions of the social shaping of technology[15] as well as the acceptance or rejection of technologies seem to be too narrow. Since what they ignore is nothing less than the fact that there are never only technologies on the one side and human beings or social groups on the other, but rather both sides are continuously transforming each other through processes of technical mediation. "We observe a process of translation", writes Latour (1991a: 116), "not one of reception, rejection, resistance, or acceptance".

While for Simondon the development of technologies appears as a "guarantee of stability" for culture (Simondon 1958: 15), Latour takes this idea a step further, arguing that it is only through technology that heterogeneous associations can be hold together. Being more stable than the fugacious and proximal interactions between humans, technologies are endowed with the capacity to translate social actions and social regulations into a more durable state. By mutually constituting and reconfiguring each other through the exchange of functions and properties, and in doing so, lending each other strength and durability, society and technology represent not separated spheres but rather "phases of the same essential action" (Latour 1991a: 129).

If technologies are opaque, that is, more of fluid geometries and embodied programs of action rather than having an essential state of being, the question becomes virulent how to distinguish technical objects and actions from non-technical ones. Once again referring to the philosophy of Simondon, Latour faces this question by understanding technology as a specific form of expression, as a certain mode of existence in the midst of other modes of being.

"It is pointless to want to define some entities and some situations as technical in opposition to others called scientific or moral, political or economic. Technology is everywhere, since the term applies to a regime of enunciation, or, to put it another way, to a mode of existence, a particular form of exploring existence, a particular form of the exploration of being – in the midst of many other." (Latour 2002: 248)

---

15 See, for example, Pinch and Bijker (1984) and MacKenzie/Wajcman (1985), as well as chapter one of this book.

And yet, even if it seems an impossible attempt to distinguish between the technical and the non-technical, at least it would be possible to characterize technology by understanding it as a movement, as a process that "liquefies all things and at the same time gives them new durability, solidity, consistency" (Latour 2013: 225). The actual *technical* element of technologies then would be the capacity to assemble and alter a multitude of different actors and forces, transforming them in a relatively solid and durable form. It is also against this background that Latour argues that we will "never find the mode of technological existence in the object itself, since it is always necessary to look beside it: first, between the object itself and the enigmatic movement of which it is only the wake; then, within the object itself, between each of the components of which it is only the temporary assemblage" (Latour 2013: 221).

As assemblages, technologies always summon a number of different actors, actions, forces, spaces, and temporalities, folding them into each other. Spatial and temporal distances become blurred just as clear distinct boundaries between the actors involved disappear. The distinction between active subject and passive object loses much of its effectiveness. Latour suggests that even in such a humble technology as a hammer, different temporalities and spaces fold into each other: eons of earth history in guise of the minerals that form the raw material for the head of the hammer, the mines of the Ruhr where the minerals came from, the wood of a centuries old Polish oak for the shaft, and so forth. Yet not only time and space fold into each other even in such a simple technology as the hammer that Latour describes but also politics, economics, and morals: the low wages for factory workers in Germany, aiming at strengthening national export capacities and in doing so causing a downward spiral of real wages throughout the EU, only to be able to sell the produced hammers at a low price on the streets of Paris, Berlin, and other places, is only one example for this. As far as Latour is concerned, nothing could hold together these specific temporalities, spaces, and actions before this specific hammer and the technical action going along with it in this specific way.

"When I grab the handle, I insert my gesture in a 'garland of time' as Michel Serres (1995) has put it, which allows me to insert myself in a variety of temporalities or time differentials, which account for (or rather imply) the relative solidity which is often associated with technical action. What is true of time holds for space as well, for this humble hammer holds in place quite heterogeneous spaces that nothing, before the technical action, could gather together." (Latour 2002: 249)

Technology, thus, not only designates an assemblage given solidity for a certain period of time but also "a sphere of politics" (Latour/Sánches-Criado 2007: 367) because it is also the political that goes into the fold. Humanity is then what emerges out of these folds, these sites of the continuous enactment of the world. Therefore, it seems less helpful to essentialistically locate humanity and human beings on the one side and technology on the other side. Rather, it becomes apparent that technology can be neither regarded as an autonomous, that is, its own impulses-following force, leaving us with no other choice than to adapt our lives to its movements and logics, nor would technologies represent neutral tools and instruments which could be used in different ways and for different purposes. But Latour goes even further, arguing that from such a perspective it could be said that "there are no masters anymore", and neither are "crazed technologies" (Latour 2002: 255). Hence, the idea of a technological extension of the natural body (Arnold Gehlen) or the concept of technology as organ projection (Ernst Kapp) would make little sense, since it is not the human that is *extended* additively through technology. Rather, humans and technologies continuously fold into each other in multiple ways, always creating something new. Latour even goes so far as to argue that the hammer gives rise to a reconfigured human being for whom, from the point at which she folds together with the hammer, a stream of new possibilities opens up that have not existed before. Far from determining human actions or even functioning as a mere backdrop for them, technologies represent what enables, authorizes, influences or prevents human actions. In contrast to Simondon's philosophy of technology, Latour, however, not only takes into account concrete political and ethical considerations but also questions of power, as I shall elaborate in what follows.

POWER AND AGENCY IN HETEROGENEOUS NETWORKS

In *Economy and Society*, Max Weber describes power as "every chance within a social relationship to assert one's will even against opposition".[16] The phrase, "every chance", already indicates that, for Weber, the source of power can be

---

16 I am following here the translation of Jürgen Habermas' *Philosophical-Political Profiles* (1983, 171) where Max Weber's famous quote has been translated much closer to the German original than in the translation edited by Guenther Roth and Claus Wittich, where "every chance" is translated as "the probability" (see Weber 1978 [1922]: 28).

manifold. In fact, "[a]ll conceivable qualities of a person and all conceivable combinations of circumstances may put him in a position to impose his will in a given situation" (Weber 1978 [1922]: 28). While power represents the property of an autonomous subject to carry out her will against the resistance of others, Weber, in the following, describes discipline as "the probability that by virtue of habituation a command will receive prompt and automatic obedience in stereotyped forms, on the part of a given group of persons" (ibid). Does that mean that nonhuman actors exercise power over human beings, if Latour speaks of processes of discipline with regard to keys and locks, speed bumps, and other technologies? While technologies do discipline us in our very actions, it would be a mistake to speak of domination here, regardless of direction. Influenced by Michel Foucault who understood power "as the multiplicity of force relations immanent in the sphere in which they operate and constitute their own organization; as the process which, through ceaseless struggles and confrontations, transforms, strengthens, or reverses them" (Foucault 1978: 92), Latour does not understand power solely as a repressive force. Rather, he tries to take into account the micro-physical relations of power that Foucault described so accurately with respect to their generative effects. Instead of limiting the scope on the sphere of social practices and thus on human beings alone, Latour argues to reconsider the question of power by involving nonhuman entities. How are dominance, order, and stability to be understood if the nonhuman is analytically included into the social web? Despite the fact that technology *does* serve as a means for the exercise of power, it would be a mistake to understand technologies primarily as instruments of dominance, as mere means to an end, since the networks within which they operate and, hence, become agentic are too complex and diffuse. Subsequently, Latour argues for extending Foucault's notion of power, allowing the taking into account of technologies in their ambivalence instead of merely as means that would support the maintenance of power in a specific way. Far from ignoring questions of power, Latour's main idea here is that one can only talk about power and power relations if one also talks about the techniques of mutual mobilizations, combinations, and transformations, since power operates always *within* and *through* heterogeneous networks composed by and of a multitude of human and nonhuman actors. Who or what acts within these entanglements is therefore never clear in advance.

Following these insights, intentional, purposeful actions are not effects of autonomous subjects that act upon their environment but rather properties of assemblages, "of institutions, of apparatuses, of what Foucault called *dispositifs*"

(Latour 1999a: 192).[17] Instead of understanding agency in a humanist sense it seems more promising here to understand agencies as collectively produced effects of heterogeneous actors evoking particular changes in the world. In this sense, it is always an assemblage, a configuration of different actors and actants entangled with each other what *acts*. An actor or actant[18] therefore is who or what introduces a difference, brings difference into the world by acting upon other agents, modifying them and their goals. Hence actors can be both individuals as well as collectives; they can be material, discursive, or figurative. To put it bluntly, there is no other way to define an actor than by its action.

Such an account however also brings with it certain difficulties as, for example, Simon Schaffer (1991: 190) illustrates, seeing Latour falling back into a hylozoism, that is, the idea that all matter would be alive. However, Schaffer overlooks that such an accusation itself rests on shaky ground since Latour exactly does not grasp intentionality as a property of individual actors – neither human nor nonhuman ones. Granting nonhuman actors agency in their very entanglements with other agents (human and nonhuman ones alike) from an *analytical* point of view does not mean to equate human beings with things on a *moral*

---

17 This idea particularly resonates with Karen Barad's understanding of agency as an enactment rather than something that someone or something has. See Barad (2007) as well as chapter three.
18 Although Latour borrows the terms "actor" and "actant" from the semiotics of Algirdas Julien Greimas, he uses them mostly interchangeably. In *The Pasteurization of France* (1988), Latour makes clear that his interest in the notions "actor", "agent", and "actant", lies primarily in their function as narrative figures. "I use 'actor,' 'agent,' or 'actant' without making any assumptions about who they may be and what properties they are endowed with. Much more general than 'character' or 'dramatis persona,' they have the key feature of being autonomous figures. Apart from this they can be anything—individual ('Peter') or collective ('the crowd'), figurative (anthropomorphic or zoomorphic) or nonfigurative ('fate')." (Latour 1983: 252, Fn. 11) In *Pandora's Hope*, Latour argues that the reason for him to prefer the term actant over that of the actor lies with respect to nonhuman entities purely in evading confusion as it may sound "uncommon" if the term actor or agent is used "in the case of nonhumans" (Latour 1999a: 180). However, in his later works an important differentiation between the notions actant and actor can be identified in the sense that an actant is understood as that which exists before it becomes an actor through a process that transforms performances in competences, actions in objects, attributes in substances, and in doing so, stabilizes an unstable entity. To put it simply, an actant is a not yet stabilized source of action that is endowed with the capacity to alter the course of events.

level.[19] If Latour is accused of putting human beings and nonhuman entities morally or even legally on the same level and, in doing so, not only misconceiving that "only humans can act", but also ignoring that nonhuman entities would "belong to another ontological region (...) not the one of spiritual world" but "the region of material nature" (Vandenberghe 2002: 52-53), not only is Latour's project understood fundamentally wrong but it is also once again spoken in the name of a "human exceptionalism" (Haraway 2008) which rests on the false premise that humanity could be detached from the "spatial and temporal web of interspecies [and technological, JB] dependencies" (ibid: 11). In fact, Latour's argument has always been that human and nonhuman actors have to be treated symmetrically within the course of action *only* with regard to their capacity to "modify a state of affairs by making a difference" (Latour 2005a: 71), but not that there would be no significant differences between a human being, a baboon, and a hammer. Only with regard to their capacity to act and thus to "make a difference in the course of some other agent's action" (ibid), human and nonhuman actors have to be treated equally, meaning symmetrically, from an *analytical* point of view. In this sense, and only in this, Latour argues, that we should not "impose a priori some spurious *asymmetry* among human intentional action and a material world of causal relations" (Latour 2005a: 76). This is necessary because if "action is limited a priori to what 'intentional', 'meaningful' humans do, it is hard so see how a hammer, a basket, a door closer, a cat, a mug, a list, or a tag could act. They might exist in the domain of 'material' 'causal' relations, but not in the 'reflexive' 'symbolic' domain of social relations" (ibid: 71).

By understanding agency as the force[20] that lies behind the continuous transformation of reality, Latour not only breaks with the difference between

---

19 John Law (1992: 383) finds even more clear words, arguing that "[t]he clarificatory point is this. We need, I think, to distinguish between ethics and sociology. The one may--indeed should--inform the other, but they are not identical. To say that there is no fundamental difference between people and objects is an analytical stance, not an ethical position. And to say this does not mean that we have to treat the people in our lives as machines. We don't have to deny them the rights, duties, or responsibilities that we usually accord to people. Indeed, we might use it to sharpen ethical questions about the special character of the human effect--as, for instance, in difficult cases such as life maintained by virtue of the technologies of intensive care".

20 Besides the fact that Latour's notion of force seems to derive from a Nietzschean lineage, Latour himself speaks relatively little about the nature of forces. In fact, it could be said that Latour keeps the term force strategically blurred, for example, when he says that he starts "with the assumption that everything is involved in a relation of

intentional, meaningful action and action as the expression for the capacity to act upon something and to put it in new pathways, but also with the dualism of action and structure. What, instead, comes into view are multiplicities of relations and entanglements of human and nonhuman entities as 'sites' from which agencies emerge. For the question of power, from this follows that it is only through the mobilization, translation, and gathering together of different actors within networks, that power relations are produced and stabilized for a certain period of time. Power only becomes efficacious within and through heterogeneous networks. What is of relevance within networks are therefore not the nodes or points but the connections and entanglements between them, as well as the mutual translations and transformations that subsequently follow. It is for this reason that within a given network no point is privileged against the other. What is more, every point itself is already a network, as Michel Callon states,

"If we wish to construct a graphical representation of a network by using sequences of points and lines, we must view each point as a network which in turn is a series of points held in place by their own relationships. The networks lend each other their force. The simplifications which make up the actor-world are a powerful means of action because each entity summons or enlists a cascade of other entities." (Callon 1986: 31)

Consequently, networks are more akin to what Deleuze and Guattari described with the botanical image of the rhizome as an alternative to the image of the tree with its linear and hierarchical lines. Instead of having a defined beginning and an end a rhizome has to be understood as an opaque network whose lines intersect and cross multiple times allowing a number of connections. Since rhizomes have no centers they unfold decentrally. "There are no points or positions in a rhizome, such as those found in a structure, tree, or root", Deleuze and Guattari (2004: 8) write. "There are only lines" (ibid). A rhizome has no beginning or end, it "may be broken, shattered at a given spot, but it will start up again on one of its old lines, or on new lines" (ibid: 10).[21] It has to be understood in this light

---

forces but that I have no idea of precisely what a force is (…) or what makes a force, for it comes in all shapes and sizes" (Latour 1988: 7, 154). Therefore, "we should not decide a-priori what the state of forces will be beforehand or what will count as a force" (ibid: 155).

21 The influence of Deleuze und Guattari on Latour's philosophy not only can be seen in the picture of the rhizomatic network but is also mentioned by Latour himself, for example, when he states in an interview that for him "Deleuze is the greatest French philosopher (along with Serres). […] I have read Deleuze very carefully and have been

that for Latour the term 'rhizome' thus "is the perfect word for network. Actor-network-theory should be called actant/rhizome ontology, as Mike Lynch says, because it is an ontology" (Latour/Crawford 1993: 263). What is important are the connections that transport and hence transform what is connected. Again, beside Deleuze and Guattari's philosophy the influence of Michel Serres can be witnessed here. Drawing on graph theory and cybernetics, Serres outlines in the *Hermès* series (1969-1980), as well as in *The Parasite* (1980), a theory of communication that breaks with the idea of the possibility of a linear transmission of information. For Serres, networks are the norm in which every point is connected with a multitude of other points. Information passes through these networks, within which it is also altered. Information, in this sense, cannot be carried through time and space without deformation but is always changed inevitably through the process of its transmission. Hence, it is clear for Serres that there is no transfer without transformation, no movement without translation.[22]

Deleuze and Guattari's cartographies of assemblages and Serres's thoughts on information and its transformation go seamlessly into Latour's understanding of networks. For Latour, just like for Deleuze and Guattari, networks do not represent transcendent, solid structures or even given things 'out there' but variable geometries, fluid ontologies that obtain stability only locally and temporarily. Networks obtain their stability only through other networks. Every "point" of a network represents a network which in turn acts upon other networks, giving them strength and stability. Hence, every point of a network also has to be considered as an actor *and* a network itself, meaning actors are always already networks themselves. Being aware of the fact that the notion of the network, as a series of nodes interconnected by lines or paths, has lost much of its original meaning and has become blunt since actual networks – in contrast to the network as an analytical figure and picture – have become ubiquitous (one has only to

---

more influenced by his work than by Foucault or Lyotard" (Latour/Crawford 1993: 263). However, while Deleuze and Guattari understood rhizomatic networks as yet to come and therefore also as a genuinely political claim, Latour assumes the facticity of them in the here and now, at least from an analytical point of view. I would like to thank Christoph Hubatschke for his helpful thoughts in discussing the differences and similarities between Deleuze and Guattari's and Latour's notion of rhizomatic networks.

22 Serres himself builds clearly on Gilbert Simondon's philosophy here. As outlined before, Simondon was skeptical of reducing information to a measurable quantity or even to a mere container for a message. For Simondon as for Serres information is always in-formation.

think of the internet, local-area networks [LANs], or broadcast networks),[23] Latour nevertheless believes in the potential of the metaphor. The reason for this lies therein that in contrast to "substance, surface, domain, and spheres that fill every centimeter of what they bind and delineate, nets, networks, and 'worknets' leave everything they don't connect simply unconnected" (Latour 2005a: 242). That is to say, for Latour, the metaphor of the network allows us to think the in-betweens.

In these in-betweens lies what Latour calls the plasma. The plasma is "that which is not yet formatted, not yet measured, not yet socialized, not yet engaged in metrological chains, and not yet covered, surveyed, mobilized, or subjectified. How big is it? [...] 'it's astronomically massive in size and range'" (Latour 2005a: 244). Latour indicates that our knowledge about this "exteriority", which is strictly speaking not an absolute exteriority since it can be mobilized, calibrated, and formatted at any time, is at best very humble. The plasma as the amorphous, as the not-yet-connected and the not-yet-formatted is the source from which the new emerges; new things, new entanglements, new associations. Just as the microbes of Louis Pasteur which had been there and yet had not been there before they were mobilized, inscribed, transformed, and thus became epistemologically, medically, and politically of weight as part of a heterogeneous assemblage consisting of Pasteur, his laboratory notes, the involved instruments and laboratories, societal and political discourses, yeast and sugar, and finally the microbes themselves, in Lille of the year 1857.

Pasteur was interested in the chemical processes behind the phenomenon of fermentation in order to be able to determine how the putrefaction of perishable foods could be delayed. Fermentation had already been known for many centuries. However, it had not been known that microorganisms were causing the putrefaction. At first, Pasteur also thought that the process of fermentation had to be a purely chemical process and that the microorganisms under his microscope only a byproduct of the fermentation or even a result of contamination. Through a number of tests in his laboratory, however, it became clear for him that it was precisely these 'contaminations' which were the main actors in the process of fermentation. Fermentation turned out to be a chemical process, however, one in which microorganisms do play a crucial role, metabolizing sugar to lactic acid. It

---

23 In a similar way also Donna Haraway problematizes the notion of the network, arguing that "[t]he U.S. military is probably one of the institutions around the world today which is the most interested in network theories. This does not mean that we no longer should use these pictures for our own work, but it should make us wonder" (Haraway 1995a: 118; trans. JB).

is against this background that Latour argues that the microbes were neither discovered somewhere 'out there' nor did Pasteur and other scientists invent them. Instead, Latour highlights the performative and generative character of (scientific) practices.[24]

"Scientific discoveries, such as the discovery of genes, depend on a configuration of machines, technologies of representation, and funds. Matters of fact are, of course, objective and real, but without their fabrication in the laboratory they would not exist. It is only retrospectively possible to say: 'The genes exist'. In the absence of the history of their research they would not exist." (Latour 2000; trans. JB)

Genes, like microbes, are real and constructed at the same time. They 'objectively' exist but it is only through specific material-semiotic enactments that they become 'real' and 'true'. Certainly, genes and microbes had existed before their fabrication, but as a matter of fact they were not the same entities as the ones that came into being with their (technoscientific) enactment. What is more, for Latour, the microbes themselves were involved in the process of their materialization and eventually became a fact, meaning true, in the moment the practice of sterilization and with it a specific idea of hygiene and cleanliness spread to hospitals and subsequently to the homes and the lives of large parts of the population of Europe. Consequently, not only the microbes have undergone a change but also the question of what counts as nature/natural and what as culture. The role of microorganisms in causing infectious diseases became increasingly apparent. The first immunizing vaccines against smallpox and other diseases were developed. A new biopolitics emerged centering around questions of hygiene and public health, eventually giving rise to eugenics as well as racist, and anti-Semitic ideas which heavily built on metaphors and images of a 'National body' (*Volkskörper*) that has to be kept 'healthy' and defended against its 'degeneration' from within as well as from the intrusions from 'the outside'. What Latour illustrates through his reading of Louis Pasteur's laboratory notes is therefore not only that the question concerning the nature of scientific knowledge production has to be reconsidered philosophically – this time however including technical objects and nonhuman entities as actors that matter[25] – but also that Pasteur's

---

24 This resonates with Karen Barad's posthumanist concept of performativity that I will discuss in the following chapter.

25 In a similar way, it could be said that Ludwik Fleck (1935; 1979) had already highlighted that the modern concept of syphilis, as an infectious disease caused by pathogenic microorganisms instead of the result of immoral behavior, was only made

laboratory has to be reframed as 'a site' where not only the microbes but consequently also the societies of the late 19th century were crucially reconfigured. It is in this very sense that Latour regards laboratories as sites where constantly discursive and material transformations, redefinitions, and reconfigurations occur.[26] Technologies such as microscopes, thermometers, and petri dishes, considered as inscription devices, had been integral parts of the enactment and materialization of the Lactobacillus. From this follows that laboratory work has to be regarded as a transformation of the world into propositions and texts with the help of technology. It is for this reason that, according to Latour, the emergence of the new (new entities, new knowledge) cannot be understood as a mere process of social construction, but also have to be considered as a material process since objects of knowledge are brought into existence through specific material practices. Objects of knowledge are, thus, neither discovered nor invented. As objects of knowledge have no inherent essences they are never once and for all determined. Furthermore, since the objects of knowledge also always have a share in the processes of their own materialization they cannot be understood as passive objects waiting to be made meaningful through social and cultural inscriptions. Rather, they represent quasi-objects, a term Latour borrows again from Michel Serres who defines the quasi-object in *The Parasite* (1982: 225) as "not an object, but it is one nevertheless, since it is not a subject". Latour employs the term quasi-object to think beyond the dualism of nature and culture, object and subject, the natural and the artificial. As hybrids without a static and stable ontology, quasi-objects are real, constructed, and discursive at the same time. Since they neither belong to the realm of the social nor to the realm of the natural, quasi-objects are not only what lies between these poles after they had

---

    possible with the help of increasingly accurate instruments such as (electron) microscopes. In this sense, technology was a relevant actor not only in shifting the meaning of syphilis but also its very ontology.

26 Latour sees a crucial disruption between laboratories of the 18th and 19th century and those of the late 20th century. From a historical point of view laboratories served as places where nature had been brought into a controlled and closed environment. The walls of the laboratory guaranteed that the unknown and uncontrollable could not make its way out into society, if something went wrong. In the age of technoscience, however, society itself became a gigantic open laboratory and experiments collective experiments in which we all take part as phenomena such as Mad Cow Disease and climate change painfully remind us.

been separated but also what forms the collective by tying heterogeneous entities together.[27]

Taking into account these considerations of Latour, it becomes evident that the call to "follow the actors" does not designate the practice of describing the world a-politically and even less can it be regarded as an uncritical empiricism but rather has to be considered as the plea for tracing the very processes through which humans and nonhumans together enact and stabilize what counts as real, alongside with its social, political, and ethical consequences.

RELATIONAL ONTOLOGY AND THE QUESTION OF THE POLITICAL

It is hard to deny that Latour is not very much interested in concrete political struggles or in the critique of prevalent social inequalities. It also seems that Latour indeed "tended to be on sites where women are absent", as Judy Wajcman (1991: 23-24) puts it aptly. Similarly, Sandra Harding argues that Latour's philosophy raises important questions, delivers some useful insights for rethinking science and knowledge production epistemologically, and in doing so "overlaps with feminist and postcolonial concerns" (Harding 2008: 45). At the same time, however, Latour would seem to show no interest in dealing with the insights of postcolonial and feminist theories, and consequently with questions of social inequality and oppression. Thomas Lemke goes even further, problematizing not only Latour's notion of the political but also his "lack of sensibility for questions of power" (Lemke 2013: 80; trans. JB). In paying little attention to the question how "social relationships to nature are inextricably bound up with capitalism, racism, and sexism, and how exploitation and dominance could be analyzed and criticized on the basis of a symmetrically operating theory" (ibid: 78), Latour would make himself an accomplice to social power relations. Andrea Whittle and André Spicer even identify an ontological realism, epistemological positivism, and political conservatism at the heart of Latour's philosophy, inevitably leading it "to legitimize hegemonic power relations, ignore relations of oppression and sidestep any normative assessment of existing organizational form" (Whittle/Spicer 2008: 623). Even though I largely agree with these critics, the accusation that Latour's philosophy would only care for an allegedly neutral and politically affirmative description of the world *as it is*, misses important insights.

---

27 See Latour (1993: 51-55).

Latour makes clear that the fact that he refuses "to explain the closure of a controversy by its consequences" would not mean that he is "indifferent to the possibility of judgement" but only that he "refuse[s] to accept judgements that transcend the situation" (Latour 1991a: 130). Taking in such a perspective means to understand truth and interest as "properties of networks, not of statements. Domination is an effect not a cause. In order to make a diagnosis or a decision about the absurdity, the danger, the amorality, or the unrealism of an innovation, one must first describe the network. If the capability of making judgements gives up its vain appeals to transcendence, it loses none of its acuity" (ibid).

Accusing Latour of apoliticism and moral relativism thus misses the point. Instead of bracketing out the political or ignoring questions of power, Latour locates political action where humans and things 'meet', where certainties break away and new associations emerge. The word 'things' here does not mean objects (*Gegenstände*) in a Heideggerian understanding as being detached from political issues but rather, following the etymological meaning of the word, a *Thing*. That is, an arena which gathers together a number of heterogeneous actors, discoursers, and forces. Things, in Latour's understanding, are always entities "out there" and issues "*in* there", and "at any rate, a *gathering*" (Latour 2004b: 233). Things are assemblies, entangled (political) issues. Understanding things as arenas for controversies to settle allows for the taking into account of them as both as facts and as values, as real and fabricated at the same time. Taking in such a standpoint opens up a perspective on things as always *disputed* things. It is in this sense that, for Latour, reality is not defined by facts but rather understood as the movement, or perhaps more figuratively, the metamorphosis from mere objects to disputed things.

What is true for technologies, namely, that they are always 'political', is thus even more true for disputed things. Disputed things are clearly 'political', "but by other means" (Latour 2007a: 813). However, the problem arises here that "[s]ince by now 'everything is political', the adjective 'political' suffers the same fate as the adjective 'social': in being extending everywhere, they have both become meaningless" (Latour 2007a: 812). To find a way to deal with this dilemma Latour suggests reconsidering politics as those discourses and practices that circle around things in the sense mentioned above. Only then, things could be brought back into philosophy, and especially into political philosophy that according to Latour is characterized by "a strong object-avoidance tendency" (Latour 2005b: 15), this time however as disputed things. Aligning himself with John Dewey's pragmatism rather than Richard Rorty or Jürgen Habermas who according to Latour would make the mistake of overemphasizing "the role of humans sitting at a table speaking with a rational basis and having a nice key-composition" (Latour/Sánches-Criado 2007: 368), and in doing so limiting democracy to the human sphere, Latour emphasizes the need for a more inclusive notion of democracy. If agency is not something that someone (the autonomous subject) possesses but rather an effect of a vast number of entangled entities and forces, politics and democracy cannot turn around human beings alone.

Through reconsidering the notion of politics, Latour ultimately arrives at a fluid materialism with a relational ontology at its heart. Precisely because such an account reframes politics as a powerful (material) activity that does not only reveal the world as it is but continuously enacts it differently, it is anything but politically indifferent. Going even a step further, science studies scholars such as John Law, Steve Woolgar, and Annemarie Mol have emphasized that such an account would abandon the singularity of reality in favor of the idea of multiple, ontologically different realities. As Annemarie Mol stresses, multiplicity is neither about perspectivism, in the sense of taking in different perspectives on the essentially same object, nor is it about pluralism. While perspectivism only "multiplies the eyes of the beholder of knowledge" (Mol 1999: 76), the adjective 'multiple' denotes the idea that different versions of an object come to existence through different practices, with the consequence that the object of knowledge is understood as being simultaneously part of different regimes and spatialities – which might coexist with one another in a productive tension or interfere with one another. "If we think performatively", John Law writes, "then reality is not assumed to be independent, priori, definite, singular or coherent. If reality *appears* (as it usually does) to be independent, prior, definite, singular or coherent then this is because it is being *done* that way" (Law 2012: 156). What Law implies here is nothing less than the idea that realities "are done" in practices. Hence, the key idea is not only that reality is enacted through (material-semiotic) practices but also that different practices enact different realties. If realities can be done, then, it is no big step to argue that they also can be undone or done differently. As a consequence, ontology becomes fluid rather than having an essence and being fixed and predetermined or simply given. Moreover, ontology becomes political insofar as different ontologies have different ethical and political consequences.[28] Steve Woolgar and Javier Lezaun highlight the possible implications of such an account, arguing that "a world of multiple realities, fluid and diverse in its ontological possibilities, is one where political questions acquire a new salience. Attending to ontological matters 'washes away the

---

28 I shall take up this idea of reality as the effect of intra-acting material-discursive entanglements of human and nonhuman entities rather than something already given and pre-existing, in the last chapter.

singularity of the real' and ushers in a pluralism that does not simply reflect a plurality of worldviews, but a plurality of worlds" (Woolgar/Lezaun 2013: 326). Instead of a plurality of worldviews, in the sense of competing interpretations and images of one and the same world, which then would only be made intelligible differently, the idea of multiple, coexisting realities intersecting, intensifying, or rendering each other (im-)possible brings into the foreground the important insight that differently enacted realities have also always different ontological, material, and political consequences. Instead of referring to an alleged essence (of the world, Nature, the body, and so forth), ontologies (in the plural) are fragile and always fluid. No longer the question which perspective on reality is the more adequate one but the inherently political and ethical question 'which of the enacted worlds are the more liveable ones'[29] and at what costs can they be realized – meaning, what are the constitutive exclusions inherent to the enactments in question – enters center stage. In fundamentally breaking with essential dichotomies such as world and words, ontology and epistemology, real and constructed, as well as with the idea of self-contained entities reduced to their alleged essences, such a relational, variable ontology displays an explicit political approach. 'Nature' loses its power to function as a point of reference for the hierarchical organization of the world, and yet it is precisely not the case that everything would be liquefied into mere *social* constructions. Moreover, in an important sense, with such an approach also the naturalization and legitimization of hierarchical power relations and dominance, of racisms and sexism loses ground. Even though Latour is not mentioning this as an effect of such a relational, fluid perspective on nature and culture explicitly, he stills seems to be aware of it, as he concludes that,

"if 'nature' is what makes it possible to recapitulate the hierarchy of beings in a single ordered series, political ecology is always manifested, in practice, by the destruction of the idea of nature. A snail can block a dam the Gulf Stream can turn up missing; a slag heap can become a biological preserve; an earthworm can transform the land in the Amazon region into concrete. Nothing can line up beings any longer by order of importance. When the most frenetic of the ecologists cry out, quaking: 'Nature is going to die,' they do not know how right they are. Thank God, nature is going to die. Yes, the great Pan is dead.

---

29 Donna Haraway's question, how "in a world full of so many urgent ecological and political crisis […] can I care" (2003b: 61), highlights precisely this point by foregrounding the urgency of thinking questions of epistemology, ontology, politics and ethics together, in order to get to "more livable worlds". I shall come back to this plea in chapter three.

After the death of God and the death of man, nature, too, had to give up the ghost. It was time: we were about to be unable to engage in politics any more at all." (Latour 2004c: 25-26)

The death of Nature finally arrives but not quite in the sense Carolyn Merchant (1980) described it against the backdrop of the industrial pollution of the environment and the exploitation of the generative forces of life itself, but rather as a certain Western idea, as an outcome of the modernist settlement, as a specific historically dateable social construction which became a commonplace (see Haraway 1991a).[30] Thus, 'we' should not be worried about the disappearance of *this* Nature as for all too long its primary function has been to support the implementation of a specific form of politics that drew its power from images and imaginations about 'naturalness' and an alleged natural order. Consequently, it can be said that Latour's considerations focus precisely on those material-semiotic practices through which differently re(con)figured realities are constantly enacted. However, since these enactments are not the effects of the actions of autonomous subjects, simply because, as mentioned earlier, agency has to be regarded as an effect of associations rather than subjects, they do not represent mere social processes.

It is for this very reason that Latour raises the question for the necessity of bringing "the sword of criticism to criticism itself" (Latour 2004b: 227). Rather than getting rid of criticism or to silence it once and for all, Latour is interested in the question how to "become critical again", how to produce "more ideas than we have received, inheriting from a prestigious critical tradition but not letting it die away, or 'dropping into quiescence' like a piano no longer struck" (ibid: 248). He concludes that this could only be achieved if criticism is understood as a "multiplication" rather than a "subtraction". Having this in mind, Latour does not want to fall back behind, or even get rid of, the insights of the Frankfurt School, feminist and postcolonial theories, or social constructionism, and "become reactionary", but rather to renew "what it means to be a constructivist" in a changed world (Latour 2004b: 246). Calling into question the usefulness of

---

30 Sandra Harding makes a similar argument, highlighting that for the majority of the people the insight that we have never been modern is anything but "big news" since the 'we' in 'we have never been modern' clearly refers to bourgeois, Western men; everyone else has indeed never been modern. In fact women, people of color, animals have always been the necessary "prerequisite for Western bourgeois men's illusions of modernity" (Harding 2008: 54).

critique as it is understood in many critical theories[31] today seems to be the only way for Latour to be able to epistemologically and politically engage with contemporary problems in a more productive way. Latour's argument here is not only that critical theories would have missed the right moment to find new weapons in order to be able to engage productively with contemporary problems but also that their very weapons had been taken over by reactionary forces. Latour claims that until recently far right and religious fundamentalists had argued absolutistic by referring to alleged natural laws or the laws of God. Today, however, it seems that right wing populists and reactionary forces are more and more highlighting the socially constructed nature of knowledge, the claim that objectivity is not possible, that science and politics cannot be separated from one another, and that facts are produced, in order to combat both science and politics that could benefit 'us' all. The idea that science equals culture and politics has become a new weapon for climate-change deniers, creationists, and neoliberal enemies of the welfare state, who are now using some of the very arguments developed by the academic left in their fight against a totalitarian understanding of science. The danger today indeed seems no longer to come solely from an excessive trust in a certain grammar of truth of the natural sciences, assuring us that all of its knowledge would represent objective and incontestable facts about the world "out there", but also "from an excessive *distrust* of good matters of fact disguised as bad ideological biases" (Latour 2004b: 227); and thus from a particular constructionist stance that liquefies materialities and facticities, making it difficult to take over responsibility and accountability for what matters and what is excluded from mattering. Instead of drawing caricatures of one another, as it happened during the so-called Science Wars, the[32] urge today lies in under-

---

31 Under the term *critical theory* Latour subsumes in a rather generalizing way all those sociological and philosophical approaches focusing primarily on questions of power and dominance as well as their critique. In some places (see, for example, Latour 2004b: 228-229), Latour also applies the term for contemporary French sociology, especially the one in the tradition of Pierre Bourdieu's work. However, if Latour draws an – admittedly questionable – analogy between contemporary forms of social critique on the one side and conspiracy theories on the other with regard to their narratives and lines of argument, it is important to understand that Latour does not accuse Bourdieu himself but rather a "too hasty" and deterministic reading of his sociology that has become en vogue.

32 The Science Wars were a debate between scientific realists (mainly physicists and mathematicians) and postmodern philosophers (primarily poststructuralist theorists, feminist philosophers, and science studies scholars) about the nature of scientific

standing that the real enemy of both scientists and scholars in the humanities and social sciences "are the religious fundamentalists who reject all knowledge that challenges their faith and the free-market fundamentalists whose policies will surely scorch the earth", as Michael Bérubé (2011: 74) puts it in a nutshell.

Precisely for this reason, Latour's goal "was never to get *away* from facts but *closer* to them" (Latour 2004b: 231);[33] not to discredit science but to

> knowledge and consequently about the nature of objectivity that started in the early 1990s and culminated in the publication of a hoax article by the physicist Alan Sokal in the journal *Social Text* in 1995. In his paper, Sokal argued that gravity would be social construct and that the newest findings in quantum physics would prove this. The article was drawing on obviously misunderstood theories of physics and mathematics and decorated with quotes from famous postmodern philosophers but nevertheless accepted by *Social Text*. In the wake of this debate, however, not only many of the scholars subsumed under the term 'postmodern philosophers' had been wrongly accused of believing that there would be no physical world per se and no way to get to objective, that is to say, situated knowledge but also critical admirers of science such as Donna Haraway, Bruno Latour, and many others had been thrown in the same pot as the enemies of science. Even though Sokal stated in a later article that his intention was to wake up the academic Left which "for the past two centuries [...] has been identified with science and against obscurantisms" but now with its turn to "epistemic relativism" would betray this heritage by undermining "the already fragile prospects for progressive social critique" (Sokal 1996), he also provided the conservative and right wing enemies of the humanities with ammunition who readily wanted to close down programs such as queer and feminist studies, critical race studies, and postcolonial studies. As a matter of fact, precisely this is happening right now at universities in Europe and the United States. The Prime Minister of Hungary, Viktor Orbán, for example, has just announced the intended shutting down of Central European University with the argument that the renewed university (since it is funded by the philanthropist George Soros, a well-known critique of the current authoritarian right-wing government in Hungary) would be a 'foreign agent' that intends to harm Hungarian society. In addition, the government announced their plan to ban Gender Studies from all universities all together, and to take back the autonomy of the Hungarian Academia of Science to decide which research projects are worth carrying out and which are not.

33 Latour considers a part of this problem as a legacy of Kant's philosophy, or even Enlightenment philosophies more generally, which were highly successful in debunking mere beliefs and irrationalities *with* the help of facts, until social constructionism as an heir to them turned the facts themselves into constructions and therefore mere beliefs,

democratize it. As Latour explains, facts gain their objectivity through specific configurations. Since there is always a multitude of chains of reference between world and words, facts are always woven into lived socio-material or natural-cultural practices. Even though facts are not mirroring reality as it is, they are not mere *social* constructions; and they certainly have real, that is to say, material consequences. The fact that what counts as knowledge and truth is in some ways constructed does not entail a relativism toward real world questions; and even less does it mean that there is no real world with real economic, social, and ecological consequences which affect the lives of myriads of humans and non-humans.

Against this backdrop, Latour's problematization of a particular understanding of critique could even be brought into line with the concern of seeing critique and critical thinking running the risk of never coming to the point of asking for the concrete political and moral constraints for the "right life in the wrong one". This is precisely one of the difficulties of Adorno's moral philosophy that Judith Butler (2012) takes up in her most recent work by raising the political question just "how to live a good life in a bad life", and "what it might mean to lead a good life in the sense of a liveable life" today. If social power relations and mechanisms of dominance, if exploitation and oppression, have to be understood to such an far-reaching extent that they invade individual contemplation toward the question of the right life and how to lead it, neutralizing it even before it can become micro-socially viral, what remains unclear is nothing less than the question of how to pursue emancipatory politics here and now. It is in this sense that Adorno (2001 [1963]: 176) argues that "the question for the good life is the question for the right form of politics", yet only to highlight in the very same sentence the impossibility of such a project by adding: "if such a right form of politics lay within the realm of what can be achieved today". Since emancipatory politics necessarily has to fail there would remain nothing left than to offer a criticism of the status quo. What becomes evident here is an imbalance between the critique of contemporary society and social power relations on the one side and theories that aim at the transformation of the status quo in a practical and emancipatory way on the other. It is also for this reason that Rosi Braidotti (2013) points out the necessity for bringing together critical theory, as an instrument for analyzing and problematizing prevalent social power relations, and emancipatory affirmative approaches that aim at the transformation of the relations and developments under critique. Critical theory has to be adapted to a

---

leaving back more or less everything as *somehow* socially constructed, from atoms and genes to black holes. See also Hacking (1999) for a very similar point.

world that has been thoroughly changed by new technologies, the technosciences,[34] and global biocapitalism.

In an important sense, however, Adorno's argument not only has to be understood as brought in position against a particular historical and political background – namely, the terrors of the NS-regime – but it must not be forgotten as well that, for Adorno, it was always clear that the question for the right life cannot be separated from the question for the right arrangement of the world.[35] If now this question is taken up once again, yet differently, by not only asking for the social arrangement (Einrichtung) of the world but also for the collective material-semiotic production or enactment (Errichtung) of the world, despite their differences, also some highly productive resonances between Adorno's and Latour's philosophy might show up. Instead of understanding nonhuman entities as mere passive resources for the production and reproduction of society, and therefore remaining within an anthropocentric logic with its constitutional exclusions, a posthumanist perspective that includes nonhuman entities analytically shifts the focus on the question of how different worlds are enacted through heterogeneous assemblages consisting of human and nonhuman actors. In doing so, not only politics is reframed as mutual enactments of differently calibrated worlds but also new perspectives and answers to current biopolitical, or rather technobiopolitical (Haraway), questions might arise. I understand Latour's

---

34 There are many stories about the origin of the notion of technoscience. Don Ihde (2010: 92), for example, traces the term even back to Martin Heidegger's philosophy of technology. The word 'technoscience' however was in fact coined by the Belgian philosopher of science Gilbert Hottois in the late 1970s. Since then the term has been taken up by many scholars from different disciplines for different purposes. In *Science in Action*, Bruno Latour, for example, mobilizes the notion of "technoscience" to break with the distinction between what counts as "science" and as "society". For Latour, we should be undecided "as to what technoscience is made of" (Latour 1987: 176). In *Modest Witness*, Donna Haraway (1997: 280) describes that her aim is to further "complicate" the notion of technoscience. For Haraway, technoscience signifies "the implosion of science and technology into each other in the past two hundred years around the world"; it designates the dense "node of human and nonhuman actors that are brought into alliance by the material, social, and semiotic technologies through which what will count as nature and as matters of fact get constituted for— and by—many millions of people"; technoscience, thus, "is heterogeneous cultural practice", it is "worldly, materialized, signifying and significant power" (ibid: 50-51).
35 See, in particular, Adorno (2005 [1974]).

image of the parliament of things as well as his politics of nature (*politiques de la nature*) precisely as an attempt in this direction.

The parliament of things represents perhaps one of the most misunderstood concepts of Latour, since it neither refers to a parliament in a metaphorical sense nor can it be even less regarded as an actual place or a place that has yet to come, granting nonhuman entities de facto the possibility to lay out their concerns. Instead, with the parliament of things Latour tries to develop an analytical instrument and political conception that allows for the taking into account of heterogeneous assemblages constituted of different actors that no longer can be separated neatly into the spheres of the natural or the social, human and nonhuman, science and politics. Transgenic animals would be an example for such entities, so is the reality of human made climate change or the hole in the ozone layer. Taking such phenomena into account, it becomes apparent that they are already subjects of political debates turning around questions of the law and ethics. Closely linked to these debates is the question who of the myriads of human and nonhuman actors are "to be taken into consideration" and "at what cost are we willing to live together well?" (Latour 2000; trans. JB). For Latour, these questions can only be answered collectively in an open, democratic process, and with them also the question "which risks we are willing to take" and "in which cosmos we want to live" (ibid).

With respect to the 'we' mentioned above inevitably the question comes up: 'Who speaks in the parliament of things'? This question, however, is a misleading one since it is neither the case that Latour's parliament of things refers to an actual or fictive gathering place where humans and nonhumans would carry out their disputes, nor should the verb 'to speak' be understood all too literally here. Rather than denoting that animals and even inanimate things would raise their voice and speak like in fairy tales, what Latour has in mind here is that humans can only speak "through mediations".

"If there were no rivers, we could not talk about rivers. There are no humans without an external world; words always mediate something that acts upon humans. [...] If people today think that chicken should be free range and riverbeds should not be rectified – then this illustrates that their words and values are in interdependence with these things. In times of intensive mass animal farming and river corrections they saw it differently. Today however we know more about living beings and this effects how we think about them." (Latour 2000; trans. JB)

What lies behind this conception is neither intersubjectivity nor representation, because similar to Deleuze[36] also Latour rejects a thinking that considers it possible to establish a correspondence between matter and discourse, world and words, nature and culture (after cutting them apart). In the wake of Deleuze who countered the idea of fixation with the idea of a continuous repetition of difference, Latour mobilizes the world of production against the world of representation, world-changing practices against mirroring repetition of a supposedly preexisting world. In Latour's philosophy representation has to be understood as "the dynamics of the collective which is re-presenting, that is, presenting again, the question of the common world, and is constantly testing the faithfulness of the reconsideration" (Latour 2004c: 248). In being entangled with nonhuman entities the more-than-human world always already reconfigures our ways of thinking and acting. Therefore, Latour sees the 'interests' of nonhuman entities as enunciated through human actions. By making nonhuman entities subjects of political, ethical, economic, and so forth, disputes, they obtain the possibility to have a say, that is, to articulate their 'interests' through humans. It is in this sense that, for Latour, nonhuman entities are also capable of formulating propositions. Articulation is a property of propositions rather than of speech: "We speak *because* the propositions of the world are themselves articulated, not the other way around. More exactly, *we are allowed to speak interestingly by what we allow to speak interestingly.*" (Latour 1999a: 144)

Latour highlights that just as there are no objective facts, there are also no objective chicken or objective rivers in whose names 'we' could speak. Therefore, the question to what extent we are talking of the 'actual' interests of, say, mad cows or laboratory animals, would not make much sense. Nevertheless, it could be asked to what extent the voices of mad cows and laboratory animals that are speaking through us are really heard and responded to? As a matter of fact, a quick look in the newspapers suggests that this can hardly be the case; otherwise there would no longer mad cows or laboratory animals, dying for the food, cosmetics, and pharmaceutical industry. On the other hand however, for Latour, the very fact that increasingly more people are talking about such things as animal rights and environmental protection laws illustrates that the 'interests' of nonhuman entities are indeed noticed, even if they are often overruled.

Arguing that Latour would ignore asymmetries of power that exist between different actors, thus, is true with regard to the fact that he does not explore social struggles as well as concrete political attempts at making acts of social

---

36 In *Negotiations*, Deleuze makes clear that what he had "been interested in are *collective creations* rather than *representations*" (Deleuze 1995: 169; italics JB).

injustice visible (which, indeed, has to be seen as a problematic blank space). On the other hand, Latour seems to be very well aware of the fact that the multitude of trails of strength also always produces winner and loser; and that for those who lose defeat could mean their complete subjugation or even annihilation. If now Latour writes against this particular background that we live in world "that no longer moves from alienation to emancipation, but from entanglement to even greater entanglement" (Latour 1999b: 30) so that the question would no longer be whether we are free or determined but "whether we are well or poorly bound" (ibid: 22), it nevertheless remains unclear what exactly being 'well' or 'poorly' bound means, and from where, meaning from which perspective, this question could be answered. Furthermore, if the 'good' ties and attachments would indeed be the "durable" ones, as Latour suggests (ibid: 23), what would that mean for the ties and attachments resulting from or drawing on racism, sexism, and other forms of social injustice that have been proven as highly durable? Even though Latour remains rather imprecise here,[37] he certainly does not have *these* things in mind when he speaks of 'good' and 'durable' ties, since he makes clear that the goal has to be to distinguish "those attachments that save from those that kill" (Latour 1999b: 30). In his more recent work, Latour gives an idea of what he is having in mind here by once again referring to the struggle against global warming in the Anthropocene (that is, our current geological period in which human activities seem to have a significant impact on the Earth's ecosystem), and the question of how to establish 'good' and 'durable' ties between humans and

---

37 Latour has been occasionally accused of being imprecise in his choice of terms and lines of arguments but also with regard to the disruptions and discontinuities in his thought (see, for example, Sokal 2008; in particular page 211). Latour, however, understands the practice of changing terms and meanings over time as an intended strategy. It is against the backdrop of this critique that he says: "The vocabulary I have used is very bad and it is meant to be bad: actant, mediation, obligatory passage point, translation, delegation, they have no meaning in themselves and they do no metaphysical work whatsoever. I never put any sort of explanatory weight on them. I don't believe the world is made of mediations, entities, or agencies. Those words are simply tools deployed to travel from one site to the next. [...] I have to warn you, my degree of reflexivity on myself is nil. I produce books, not a philosophy. Every book I am involved with is a work of writing that has its own categories and its own makeup. I cannot transform all of these books into a unified field of thought that would remain stable over time and of which one book would simply be coherent manifestations. On the other hand, I don't believe in being irresponsible for what I have written." (Latour 2003: 18-19)

nonhumans, between what might be called society and what might be called the Earth, putting forward to existential need for a cosmopolitics or even Gaia politics.[38]

THE ABSENT PRESENT BODY

As outlined above, following Latour, we can only speak about technology when we take into account the processes of mobilization and reconfiguration between human and nonhuman actors. Rather than denoting a proper object, technology refers to a process that liquefies or sets in motion and solidifies or temporality stabilizes boundaries and properties. Instead of already starting with the assumption that technology is always political, meaning that certain technologies would be inherently democratic while others authoritarian, Latour makes explicit the need to shift the focus on rhizomatic networks in order to follow the actors within them. Far from understanding technologies as once and for all in steel and concrete condensed social power relations, such a perspective demonstrates that technologies represent fluid geometries, sites where agency is distributed to a plethora of heterogeneous human and nonhuman entities, and where it is never clear in advance who or what 'acts' and with what effect.

In taking in such a perspective, one-dimensional explanations that understand technology either as a powerful force, determining the course of history, or only as a passive carrier of social meanings, in particular, power relations, are left behind. What is more, the distinction between the natural and the artificial loses much of its power. Latour stresses that the important question was never whether something is artificial or not but "whether the cosmos is composed of monsters", of technologies which are considered to be 'controllable' and for that reason not even sought to be included into the collective.[39] Rather than being

---

38 See Braidotti (2013) for a similar call for a "planetary politics" in the struggle against global warming.
39 The notion of the 'collective' is once again a term Latour borrows from the philosophy of Michel Serres. Even though Latour speaks of 'the collective' in certain places, there is no such thing as 'the collective' but only *collectives* in plural. Collectives are volatile associations, assemblies of humans and nonhuman entities and forces. Collectives are what replace society and nature as opposed poles in Latour's philosophy. Apart from the influence of Serres's philosophy, the notion of the collective group (le groupe collectif) from Gilbert Simondon (1958: 245) also shines through here. Collective groups are ensembles or specific arrangements of technical and psychological

embedded in technology itself, for Latour, this is where the danger comes from and why consequently "democracy has to search for the remaining hope at the bottom of the box [of Pandora]" (Latour 2000; trans. JB).

Against this particular backdrop, the argument that Bruno Latour's philosophy would be politically indifferent, having not much to add to feminist and other critical[40] endeavors that attempt to understand the ethical and political consequences of technologies, seems only partly true. In fact, I do believe that Latour's philosophy does not only provide us with a highly promising understanding of technology as a process rather than a substance, but also with an inherently political understanding of technologies. However, when it comes to the question of the body and embodiment, it seems that Latour appears to lack a notion of the material body. Even though Latour illustrates the impossibility of separating technologies from the human, he speaks surprisingly little about the material, lived body. The question of the body and corporeality only finds its way into Latour's philosophy through two ideas: On the one hand, the body appears as a resource for a number of organic metaphors Latour is working with when he describes both scientific knowledge production and the work of science studies scholars. On the other hand, the body comes to the fore in Latour's turn to affects. While in the first case Latour adds materiality to the work of both scientists and science studies scholars, in the second case he develops an understanding of the body as something that is always moved by other entities and forces, being eventually almost indistinguishable from those very forces and entities, but nevertheless not the same.

Be it in the discussion of Louis Pasteur's laboratory notes or in the reconstruction of Frédéric Joliot-Curie's race for the first human-made nuclear fission as a chain reaction against the backdrop of the German invasion of France, *Pandora's Hope* is traversed by organic metaphors drawing on the biological body and its 'interior'. Describing the circulation of scientific facts, Latour (1999a: 310) comes up with a quasi-organic, corporeal image of knowledge Just like in organisms, science, too, could be understood as having its own circulatory

---

components and forces. Technical activities produce these ensembles and with them what Simondon calls transindividual collectives ('collectifs trans-individuels'). For Simondon, the technical world is characterized by a sea of possible connections and rearrangements. It is precisely this idea as well as the notion of assembly that had a significant influence on the philosophy of Gilles Deleuze and Félix Guattari; in particular, on their concept of agencement (assemblage).

40 The term 'critical' refers here to both reconstructive and deconstructive practices, aiming at more liveable worlds, rather than debunking ones.

system and heart. Following this analogy, Latour highlights that the "notion of a science isolated from the rest of the society will become as meaningless as the idea of a system of arteries disconnected from the system of veins" (ibid: 80). Science here seems to become almost an animated, vibrant organism whose blood flow, that is, the ways in which facts circulate, could be described in the same way as William Harvey[41] did for organisms, meaning "blood vessel after blood vessel" in order to be able to reconstruct "the whole circulatory system of science" (Latour 1999a: 80). In contrast to the reductionist view of the 'science warriors',[42] who would only see the cut out heart "brightly lit on an operating table", and in doing so would only mutilate the circulatory system, the task of science studies scholars, as self-proclaimed admirers of the sciences, would be to treat "a bloody, throbbing, tangled mess, the entire vascularization of the collective" (ibid: 109) in order to be able to truly understand the nature of scientific knowledge production. Surely, the heart is required to pump in and out but it is not viable without the lungs, the kidneys, the vascular system, and so forth. Far from being cut off from society the more connections science has, the more real it becomes, meaning, "the more chance there is for accuracy to circulate through its many vessels" (Latour 1999a: 113-114). In describing science and scientific knowledge production by using concepts referring to living organisms and attesting to this living thing a certain kind of precariousness and vulnerability that scholars in science and technology studies ought to protect – even at the risk of evoking a war[43] – Latour, however, also draws a problematic picture of the

---

41   William Harvey was an English physician and the first to describe the blood flow on the living organism in detail at the beginning of the 17th century. See the very insightful biography written by Thomas Wright (2013).
42   In *Pandora's Hope* and beyond, Latour applies the term 'warriors' to the critics of Science Studies (see Latour 1999a).
43   It cannot and should not be denied that Latour seems to have a certain affinity to martial terms and the use of bellicose rhetorics. Terms such as wars, warriors, and battlefields can be found throughout his work. Latour goes so far as to entitle the French edition of *The Pasteurization of France* (1988), in homage to Tolstoi's novel, *War and Peace, Les microbes : guerre et paix, suivie de Irréductions*. Sometimes Latour even compares himself to a "good military officer" who must "retest the linkages between the new threats he or she has to face and the equipment and training he or she should have in order to meet them" (Latour 2004b: 231). It is for this reason that Donna Haraway not only problematizes "the structure of heroic action" in Latour's project but also argues that "[f]or the Latour of Science in Action, technoscience itself is war, the demiurge that makes and unmakes the world. The action in science-in-the-

"science warriors" as purely destructive villains, threatening the circulatory system of science in its integrity and health from within in a similar way germs and viruses would do. Not only that Latour constructs a straw man, 'the science warriors', reducing the complexity of the debate to a caricature, such an analogy is also running the risk of getting dangerously near to organicist and biologistic metaphors and pictures that flourished in political debates between the late 19th and early 20th century in Europe and the US.[44]

But there is a second dimension to talk about the body and corporeality in Latour's philosophy. Influenced by Deleuze's reading of Spinoza, Latour understands having a body as *"to learn to be affected*, meaning 'effectuated', moved, put into motion by other entities, humans or non-humans" (Latour 2004a: 205). In the wake of Deleuze's explications on the body, Latour uses the sense of smell as an example to make explicit that acquiring a body represents a process that always goes along with the ways of experiencing the world and making it intelligible. The body and "body parts are progressively acquired at the same time as 'world counter-parts' are being registered in a new way. Acquiring a body is thus a progressive enterprise that produces at once a sensory medium *and* a sensitive world" (Latour 2004a: 207). Hence, Latour argues that training the nose for the perfume industry, for example, means to be affected, that is, to be moved bodily from the influences of chemicals. Latour does not understand the ability to perceive subtle nuances of odors from a vast spectrum of smells as an effect of the training of the nose in order to be able to perceive different chemical fragrances but rather as "becoming a 'nose'", as being "affected, that is *effected* by the influence of the chemicals" (Latour 2004a: 207). It is in this sense that Latour argues that acquiring a body has to be understood as the "progressive

---

making is all trials and feats of strength, amassing of allies, forging of worlds in the strength and numbers of forced allies [...] Trials of strength decide whether a representation holds or not. Period. To compete, one must either have a counterlaboratory capable of wining in these high-stakes trials of force or give up dreams of making worlds" (Haraway 1997: 34). However, to be fair, Latour himself seems to increaseingly question the helpfulness of such terms and narratives in his more recent work.

44 Philip Sarasin (2006) points out how in the second half of the 19th century bacteriology and immunology have functioned as a reservoir for highly problematic images and metaphors such as 'foreign bodies', 'wandering germs', 'intruders', and so forth, which were understood as infiltrating and threatening an allegedly 'healthy' and 'pure' 'national body'. Eventually, many of these metaphors were employed for the discrimination and later even for the legitimization of the extermination of social groups and individuals imagined as the foreign, invading, and impure Other.

enterprise that produces at once a sensory medium and a sensitive world" (ibid). The world becomes richer in noticeable chemical fragrances. Where once a number of unrecognizable chemicals used to bombard the nose, now an ocean of new odors opens up. In a certain sense the nose would, according to Latour, not only learn to differentiate between a number of odors but is also 'brought to speak', being now able to articulate the different odors. It is for this reason that Latour takes a stance for the lived, always entangled, and embodied body that is never a mere material container for the mind or separated from the world 'out there', but on the contrary, always entangled with it. "I want to be alive", Latour writes, "and thus I want more words, more controversies, more artificial settings, more instruments, so as to become sensitive to even more differences. My kingdom for a more embodied body!" (Latour 2004a: 211-212).

Even though Latour draws on an anything but thrilling example,[45] in an important sense, he is still not only able to demonstrate with it that the boundaries between the body and the world are permeable and ever shifting (in unpredictable ways), so that neither the world nor material bodies would represent self-contained entities, but also to highlight that within his relational ontology the world and material bodies are understood as constantly reconfiguring each other in the very same movements. Therefore, however trivial or even obscure Latour's example might seem at the first glance, it also opens up a perspective that illustrates that it is impossible to talk about such things as an essence of material bodies, in the sense of a static core, or to define what the body 'actually' or even 'by nature' *is*. Instead, Latour breaks with the distinction between the world 'out there' and the world experienced by and through bodies, between how the world is and how the world is understood and becomes (bodily) intelligible, and hence also with the idea of an insurmountable gap between world and words, the material and the semiotic. Moreover, in breaking with traditional dichotomies and anthropocentrism such an approach allows us to understand that there is no technologically untouched body or nature 'out there'. As outlined earlier in this chapter, a consequence of such an understanding is that nature and supposedly natural bodies no longer have to be defended against the technological as a colonizing and objectifying force. Instead, the organic and the mechanical, material bodies and technical objects, nature and culture could be understood as cuts,[46] as

---

45 Latour (2004a: 206) seems to see this fact as an advantage, since it would be "much less dramatic than the medical cases often automatically associated with discussion about the body".

46 I am drawing on a key term of Karen Barad's philosophy here. For a detailed discussion of the notion of the agential cut, see chapter three.

sedimented moments of the same movement. This is precisely why, for Latour, "the very notion of culture is an artifact created by bracketing Nature off. Cultures – different or universal – do not exist, any more than Nature does. There are only natures-cultures" (Latour 1993: 104).[47] However, it is important to understand that this does not mean that nature and society have "to be 'maintained together' so as to study 'symmetrically' 'objects' and 'subjects', 'nonhumans' and 'humans'" (Latour 2005b: 76). Rather, what Latour means here is that natures and cultures are only to have in the plural form of the words, and never separated from each other. This idea illustrates once again what Latour means by arguing that the notion of 'the social' never explained anything since 'the social' cannot be explained out of 'the social' itself, purged from the natural. Rather, 'the social' is something that has to be explained itself in the first place and always anew. Instead of referring to a solid substance, 'the social' designates the effect of interfering relations and practices, a movement bringing together heterogeneous entities and forces.[48] That being so, 'the social' just as the body resembles much of what Deleuze (1992) describes as a gaseous composition without a clear shape or determined boundaries. In contrast to traditional sociological approaches which always already knew what 'the social' is and who would belong to it – namely, intentionally acting human beings (that is, for a long time mainly *white* bourgeois males) while everyone and everything is only to be considered as a resource for the reproduction of 'the social'[49] – in

---

47 For the term naturecultures see also Haraway (2008; in particular, chapter nine).
48 It is interesting that for more than a century the word 'social' denoted a sui generis force if one considers that the etymological roots of the word 'social' can be found in the Latin word 'sequi' which means 'to follow', 'to come'. See also Latour (2005a: 6) who argues that "[t]he Latin socius denotes a companion, an associate. From the different languages, the historical genealogy of the word 'social' is construed first as following someone, then enrolling and allying, and, lastly, having something in common. The next meaning of social is to have a share in a commercial undertaking. 'Social' as in the social contract is Rousseau's invention. 'Social' as in social problems, the social question, is a nineteenth-century innovation."
49 It would be a mistake to think that Marxist philosophy is spared from this problem. Even though, especially in the *Economic and Philosophic Manuscripts* (1844), Marx understands nature as "man's *inorganic* body", nature however, in contrast to human beings who would actively produce and pursue their own history, remains ultimately only a mute and ahistorical resource for the reproduction of society as well as the selfactualization of humanity. It is in this sense that human activity, for Marx, is the continuous engagement with nature, meaning the (technological) transformation and

beforehand, Latour foregrounds the argument that the question for the nature of the social realm as well as the question who would belong to it has to be raised always anew, and cannot be answered ahistorically and once and for all.

Consequently, such an account provides a promising starting point for contesting the idea of a colonization, denaturalization, and even dissolution of material bodies through modern technology – as I have outlined and problematized in the previous chapter. Instead of considering the body as the seat of the soul, and yet, or precisely therefore, as a mere passive and subjugated object without any kind of material agency, bodies could be understood as interfaces or sites where different entities and forces meet and recalibrate one another. Hence, the body could cease to represent a mere blank slate for social and cultural inscriptions and could become significant as an actor itself. But, despite the fact that Latour helps us to understand that the body is always a body that is entangled with other bodies, the question remains open as to how differently reconfigured bodies come to matter. As I have argued in the previous chapter, there is no such thing as *the* body, not only because material bodies are always multiple, always in-becoming, and always entangled with other bodies (human and nonhuman, organic and machinic ones), but also because just as there is no view from nowhere there is also no completely unmarked body. In fact, even before a body is born it becomes marked by sex, race, dis/ability, species, class, ethnicity, and even nationality (for example, where jus sanguinis[50] is applied to the body of the embryo). Therefore, to argue that there is no such thing as an unmarked body

---

commodification of nonhuman entities and forces through labor. This is precisely what the Frankfurt School has characterized as a 'messianic anthropocentrism'. Such a position is, ironically, not too far away from what Niklas Luhmann foregrounds in his systems theory: namely, the belief that nonhuman entities are merely the environment, meaning resource, for human beings and human society. Just in the moment ideas about the limits of growth, the extinction of species, and ecological risks and catastrophes are gaining ground and making their way into public debates, Luhmann argues that species, ecosystems, and forests cannot be "irritated" about their disappearance and death since they are mere human categories, that is, only "an invention of human judgment" (Luhmann 1997: 158). As if ecological pollution and the death of entire species would be nothing but a matter of communication between human beings and the question of responsibility and accountability in the face of the sixth mass extinction purely optional.

50 Jus sanguinis, or 'the right of blood', designates the law by which the citizenship of the newborn is determined by the citizenship of its parents (or at least of one part of them) rather than by the place of birth (jus soli).

does not mean to level fundamental differences when it comes to the question how and which marks are left on bodies along with which ethical and political consequences. In an important sense, I furthermore do not want to reduce this idea to human bodies alone, as one of the most powerful marks that can be left on bodies is the category 'species'. Bodies are always – even though in fundamentally different ways and with fundamentally different ethical and political consequences – bodies marked by race, class, sex/gender, ethnicity, nationality, dis/ability, species, and much more. However, the question remains open as to how to grasp these processes of becomings and reconfigurings, particularly when technologies and technoscientific practices come into play.

3 Re(con)figuring the Apparatus

> Reality is an active verb, and the nouns all seem to be gerunds with more appendages than an octopus. Through their reaching into each other, through their 'prehensions' or graspings, beings constitute each other and themselves. Beings do not preexist their relatings. 'Prehensions' have consequences. The world is a knot in motion.
> —Donna Haraway/*The Companion Species Manifesto*

It is no coincidence that feminist theorists, scholars of color, and those speaking from marginalized locations in particular put questions of materiality and material bodies at the center of critical analyses, given their historically long lasting identifications with nature. Queer and feminist scholars have not only deconstructed the idea of the gendered and sexed body as a matter of fate but also shifted the focus toward questions of the body and embodiment. Fighting for bodily autonomy, reproductive freedom, and sexual rights, feminist and queer scholars have turned to the body as a powerful source for emancipatory politics.[1] The concept of the apparatus of bodily production is only one outcome of these critical endeavors, however, a highly promising one for turning to questions of

---

1 For example, in *Seizing the Means of Reproduction. Entanglements of Feminism, Health and Technoscience*, Michelle Murphy follows the history of feminist struggles for reproductive freedom in California during the 1970s and 1980s. Murphy reconstructs how feminists appropriated, revised, and in doing so (re)politicized biomedical technologies, methods, and practices. Analogous to Marxist calls to seize the means of production, for Murphy, these practices can be understood as feminist attempts at seizing the means of reproduction in order to craft responsible "politics out of technoscience and bodies, rather than, say, labor and citizenship" (Murphy 2012: 2-3).

biology, technology, politics, power, and agency in their entanglements with one another.

The notion of the apparatus of bodily production can be traced back to feminist literary scholar Katie King (1991) who employed her figure of the *apparatus of literary production* to shed light on the question of how literature comes into existence at the crossroads of global capitalism, art, and technology. In King's account literature is produced through apparatuses of literary production where language itself represents as much an actor as the author herself. Consequently, for King, there is no author in a traditional humanist sense any more; rather, it could be said that the authorship always lies in the 'hands' of a collective of human and nonhuman, material and textual entities.[2]

ON MATERIAL-SEMIOTIC ACTORS AND GENERATIVE NODES

In her seminal essay, *Situated Knowledges* (1988), Donna Haraway takes up King's figure of the apparatus of literary production and reframes it as a tool for feminist analyses of technoscientific practices, highlighting the objects of knowledge as active entities. Problematizing a particular notion of scientific objectivity, namely, the so called "god trick" of seeing everything from nowhere, which rests on the belief that an object of knowledge is "a passive and inert thing" (Haraway 1991a: 197), Haraway emphasizes that no matter how mediated the world may be, it is never a mere object. For Haraway it is only through a specific "analytical tradition" that everything gets turned into a resource for appropriation, leaving back the objects of knowledge as mere "matter for the seminal power, the act, of the knower" (ibid). Rather than being passive matter, the objects of knowledge are always actively involved in the reconfiguring of the world. Against the backdrop of this idea, the concept of the apparatus of bodily production emphasizes that bodies and organisms are not born but rather

---

2   King's concept thus can be, at least partially, regarded as staying in the tradition of a poststructuralist line that leaves behind the idea of individual authorship. After Roland Barthes called out the 'death of the author' in 1967, contesting a reading that assesses the meaning and significance of a text widely on the basis of the biography of the author as an embodied subject, Michel Foucault took up this idea in his talk "Qu'est-ce qu'un auteur?" at the Société française de philosophie two years later, arguing that it would be a mistake to believe that the author precedes her work; rather, both are mutually co-constituting one another.

produced "in world-changing techno-scientific practices by particular collective actors in particular times and places" (Haraway 1992: 297). However, this is not to say that bodies and organisms are mere *social* constructions Referring to biology, Haraway describes apparatuses of bodily production as material-semiotic arrangements of human and nonhuman entities as well as practices through which organisms are produced. Similar to Bruno Latour's (1993) reading of Louis Pasteur's laboratory notes, in which Latour describes the formation of lactic acid bacteria as a material-semiotic process in which the bacteria themselves were actively involved, Haraway argues that bodies and organisms are neither epistemologically nor ontologically preexisting entities, waiting somewhere 'out there' to be discovered, and can even less be regarded as mere end products of processes of social construction. Rather, bodies and their boundaries materialize through particular practices, that is, through particular apparatuses of bodily production. For Haraway, thus, bodies (human and nonhuman ones alike) represent "active participants in the constitution of what may count as scientific knowledge" (Haraway 1989: 310). It is only through particular apparatuses of bodily production that "boundaries take shape and categories sediment", as Haraway (1994: 64) highlights. As a concept that stays for the generative powers of material-discursive practices, the notion of the apparatus of bodily production hence shifts the view from the politics of representing *and* intervening (Hacking) to the politics of representing *as* intervening.

Inseparably tied to the notion of the apparatus of bodily production is the concept of the material-semiotic actor. Haraway's material-semiotic actor highlights that objects of knowledge are active parts, that is, as generative nodes of the apparatuses of bodily production. In the concept of the material-semiotic actor materiality and metaphor merges together. Being very well aware of the fact that poetry is not epistemology and even less biology, Haraway is nevertheless convinced that the concept of the material-semiotic actor echoes the collapse of the supposedly clear distinction between the material and the textual, the grown and the made, active subject and passive object, and hence also between epistemology and ontology, yet without conflating them.

"Are biological bodies 'produced' or 'generated' in the same strong sense as poems? From the early stirrings of Romanticism in the late eighteenth century, many poets and biologists have believed that poetry and organisms are siblings. Frankenstein may be read as a meditation on this proposition. I continue to believe in this potent proposition, but in a postmodern and not a Romantic manner of belief I wish to translate the ideological dimensions of 'facticity' and 'the organic' into a cumbersome entity called a 'material-semiotic actor'." (Haraway 1991a: 200)

The figure of the cyborg is perhaps the most famous and at the same time also most controversial example of Haraway's figural realism. As a material-semiotic figure, Haraway's cyborg fuses together feminist fiction and material reality, opsitional thinking toward more liveable worlds and the implications of the military-industrial complex and neoliberal economy with its devastating effects of women in the Global South. Haraway's cyborg is at once both a metaphor for fragmented identities and "our ontology" (Haraway 1991a: 150), meaning, our bodily and social reality in the late 19th century. In any case the cyborg is what "gives us our politics" (ibid).

The cyborg is a product of Cold War politics. As a material-semiotic actor the cyborg emerges as a central figure in Haraway's 'Cyborg Manifesto' against the backdrop of the Reagan era in the US. Having been asked by the *Socialist Review* to write a short paper about the consequences of the re-election of Ronald Reagan for the future of feminist and socialist politics in 1984, Haraway eventually ended up with what became known as the "Cyborg Manifesto". Neither the title, which clearly refers to Marx's *Communist Manifesto*, nor the choice of the text type itself represents a coincidence. In an important sense, Haraway's "Cyborg Manifesto" can be regarded as an attempt at queering the traditionally masculinely coded text type of the manifesto. As Karin Harrasser (2006) points out, women are absent as political actors in almost every historically influential manifesto, be it Marx and Engels's Communist Manifesto, the Dada Manifesto, or Marinetti's fascist Futurist Manifesto. While Marx and the Dadaists are talking little about actual women and even less about gender relations, for Marinetti women seem only to be present as a projection screen for masculinist fears and misogyny. In emphasizing the connectivity and solidarity with animals and even machines, Haraway thoroughly queers a traditionally masculinist and anthropocentric text type. Haraway's cyborg is thus about "transgressed boundaries, potent fusions, and dangerous possibilities which progressive people might explore as one part of needed political work" (Haraway 1991a: 154).

On the other side, however, the cyborg, in Haraway's understanding also represents "the illegitimate offspring of militarism and patriarchal capitalism" (Haraway 1991a: 151), a weapon in the space-race against the Soviet Union. As a cybernetic organism, *this* cyborg dates back to the research of the psychologist Manfred E. Clynes who worked on computerized feedback control of biological systems and the clinical psychiatrist Nathan S. Kline who was working as director at the Rockland State Hospital in New York in the 1950s and 1960s. Both Clynes and Kline were exploring the human nervous system in an attempt at digitally mapping it using highly sophisticated computer systems before they wrote the essay "Cyborgs and Space" for the journal *Astronautics* in 1960. The article itself built on a talk entitled "Drugs, Space, and Cybernetics: Evolution to Cyborgs" Clynes and Kline gave a year earlier at the Air Force School of Aviation Medicine in San Antonio, Texas. In their article, Clynes and Kline were raising the question how to alter human bodily functions in order "to meet the requirements of extraterrestrial environments" (Clynes/Kline 1960: 26). For both it was clear that it would be more logical to adapt human bodies to particular conditions one might find in outer space and even on alien planets than to provide an earthly environment for astronauts in space.

"If a fish wished to live on land, it could not readily do so. If, however, a particularly intelligent and resourceful fish could be found, who had studied a good deal of biochemistry and physiology, was a master engineer and cyberneticist, and had excellent lab facilities available to him, this fish could conceivably have the ability to design an instrument which would allow him to live on land and breathe air quite readily." (Clynes/Kline 1960: 26)

Drawing an analogy between the fish in the aforementioned thought experiment and actual human beings in space, Clynes and Kline argued that with the help of modern technologies such as self-regulating human-machine systems that stabilize bodily functions automatically, human beings could adapt to the conditions of outer space. Implants would automatically control bodily functions, lower, for example, the heartbeat and body temperature, if necessary, or continuously inject biochemical substances against cosmic radiation, and in doing so would it even made needless to breath air since oxygen would be synthesized or brought directly into the bloodstream through specific implants. Neither being a slave to the own metabolism that has evolutionary adapted to the conditions on Earth, nor being a "slave to the machine", human beings would be freed to explore the universe.

"If man in space, in addition to flying his vehicle, must continuously be checking on things and making adjustments merely in order to keep himself alive, he becomes a slave to the machine. The purpose of the Cyborg, as well as his own homeostatic systems, is to provide an organizational system in which such robot-like problems are taken care of automatically and unconsciously, leaving man free to explore, to create, to think, and to feel." (Clynes/Kline 1960: 27)

The cybernetic organism as the self-regulatory coupling of biological body and mechanical machine was born. "For the exogenously extended organizational complex functioning as an integrated system unconsciously, we propose the name 'Cyborg'", Clynes/Kline (1960: 27) write. Noting that "there are references in the Soviet technical literature to research in many of these same areas" (ibid: 76), Clynes and Kline reinforced the urge for boosting military research on cybernetic organisms if the US ought to beat the Soviet Union in the space race.

The first real cyborg, in the sense of a self-regulatory organism-machine-system, was a small rat which had an osmotic pump surgically attached to its tail that continuously injected chemicals into the body, establishing a feedback loop between the metabolism of the rat and the pump that automatically dispensed chemicals into the rat's body as they were needed. Ironically, in an interview many years later Clynes insists that the cyborg is fully human (or, for that matter, animal) and that the idea was never to erase 'the human' but to "make it possible to exist *qua* man, as man, not changing his nature, his human nature that evolved here" (Clynes/Gray 1995: 47). Clynes is very specific in emphasizing that what he had in mind was *enhancing* the natural human body through prosthetic organs, and not getting rid of it. For Clynes, it seems, the goal was never to overcome human nature but rather to alter it "for the maintenance of the person. It wasn't changing their nature" (ibid: 48). What becomes apparent here is the Cartesian nature of Clynes and Kline's cyborg concept. Ian Hacking states that, especially in Clynes's account,[3] the feedback loop established between the body on one side and machines and chemicals on the other only aims at altering the body and its function in order to set bodies free for the exploration of hostile environments, while the mind and with it 'human nature' is left "as it is" (Hacking 1998: 209). Consequently, not only the dichotomy of mind and body remains intact hereby but also the dichotomy of organic and mechanical, of active creative mind and passive brute matter, is upheld.

Haraway's cyborg, however, was always much more than only an illegitimate offspring of militarism and 1960s Cold War space race – a fact that all too often got lost in the reception of her work. As a material-semiotic figure, the cyborg has also always stood for the implosion of fundamental dichotomies. And yet even in Haraway's account it cannot be denied that the figure of the cyborg, particularly with the idea of the fusion of the organic and the inorganic, the biological and the mechanical, body and technology, implies that it is only relatively recently, namely through modern science and technology, that once seemingly substantially distinct spheres have become blurred. Haraway's cyborg is

---

3   Hacking notes within this context that "it was probably Clynes who was keen on leaving the mind as it is, and who was Cartesian, while Kline was into mind control with a vengeance. But the cyborg was Clynes's idea" (Hacking 1998: 210).

clearly about a specific mid or late 20th century political and technoscientific constellation. In a certain sense, Haraway's cyborg could even be read as implying that a once natural, meaning, technologically untouched body is becoming more and more technologized, making it less and less possible to maintain the idea of a clear distinction between the born and the made. And indeed, the Derridean feminist Vicki Kirby notes that,

"Given the many disclaimers to the contrary it is ironic that the cyborg is perhaps the most recent of Cartesian recuperations. Haraway's insistence that '[t]he cyborg skips the step of original unity' forgets that it is against the unity of 'the before,' the purity of identity prior its corruption, that the cyborg's unique and complex hybridity is defined." (Kirby 1997: 147)

At least partially, Haraway's cyborg remains caught in the logic of "the calculus of one plus one, the logic wherein preexisting identities are *then* conjoined and melded" (Kirby 1997: 147). Hence, it seems that it is indeed the case that, "even if only retrospectively", as Kirby (ibid) states, the natural and the artificial 'parts' of the cyborg can be separated from each other. This becomes, for example, evident when Haraway asks "why should our bodies end at the skin, or include at best other beings encapsulated by skin" (Haraway 1991a: 178), but at the same time also believes that "[c]ommunications technologies and biotechnologies are the crucial tools recrafting our bodies" (ibid: 164). In another place, Haraway even writes that "cyborgs are about specific historical machines and people in *interaction*" (Haraway 1997: 51; italics JB). While it is without doubt true that certain technologies can have far-reaching effects on the body and its functions, within Haraway's figure of the cyborg technology remains partially something external to the body, something that only cumulatively extends the body. It seems that both natural organic bodies and artificial technologies precede their connection as two distinct spheres. With the figure of the cyborg, Haraway, thus, reiterates the very categorial dichotomy she wants to overcome at least to a certain extent. If the cyborg is really our ontology and if it therefore also gives us our politics, as Haraway suggests, then, it often seems that this ontology and politics is one of *connection* rather than of *entanglement*.

Kirby is therefore right with her argument that Haraway's cyborg at least partially remains caught in the logic of Cartesian distinctions.[4] In an important

---

4   This seems to be true for Haraway's earlier work. On the other hand, it is worth noting that even in the *Cyborg Manifesto* also passages can be found where Haraway writes, for example, that "[t]he machine is not an it to be animated, worshipped, and

sense, however, Haraway's figure of the cyborg was not so much about propagating the actual hybridization of what was once organic and natural with artificial and technological components, as it was meant to describe how, at the end of the 20th century, "particular sorts of breached boundaries confuse a specific historical people's stories about what counts as distinct categories crucial to that culture's natural-technical evolutionary narratives" (Haraway 1995b: xvi).

Nevertheless, in her later work, Haraway herself became increasingly unhappy with the figure of the cyborg, stressing that "as an oppositional figure the cyborg has a rather short half-life" (Haraway 2003a: 54). While the initial task of the cyborg figuration was "to do feminist work in Reagan's Star Wars times of the mid-1980s", Haraway admits that "[b]y the end of the millennium, cyborgs could no longer do the work of a proper herding dog to gather up the threads needed for critical inquiry" (ibid: 60), highlighting the need for new promising figures such as the coyote and the companion species;[5] even though none of

---

    dominated. *The machine is us, our processes, an aspect of our embodiment*" (Haraway 1991a: 180; italics JB). I consider such a reductionist understanding of the cyborg, thus, more of a problem in the work of Chris Hables Gray than in Haraway's work. In Gray, the figure of the cyborg seems to lose much of its emancipatory, oppositional function by being all too narrowly limited to the actual "melding of the organic and the machinic, or the engineering of a union between separate organic systems" (Gray 1995: 2). In this sense, for Gray, the cyborg denotes the "merging of the evolved and the developed" (ibid). For Gray, cyborgs are "everywhere", proliferating in popular culture as much as in medicine, leading him ultimately to the conclusion that almost everyone of 'us' would be a cyborg by now: "Anyone with an artificial organ, limb or supplement (like a pacemaker), anyone reprogrammed to resist disease (immunized) or drugged to think/behave/feel better (psychopharmacology) is technologically a cyborg. [...] It's not just Robocop, it is our grandmother with a pacemaker. [...] arguably anyone whose immune system has been programmed through vaccination to recognize and kill the polio virus." (Gray 1995: 2-3) In my reading, also Adele Clark too strongly emphasizes the prosthetic aspects of the cyborg, identifying it primarily as a body coupled with technological devises. It is in this sense that Clarke also regards the "postmodern lived reproductive body" as a "transformed, manipulated, customized" [cyborg, JB] body (Clarke 1995).

5   Haraway emphasizes that the figure of the companion species gives her "another way not just to think about kin groups of feminist figurations, but also to live them. Feminists, anti-racists, and socialists have always argued for collective action if we are to have any hope for more livable worlds. In that sense, like the 'Cyborg Manifesto,' the move to a 'Companion Species Manifesto' is an effort to do socialist feminist anti-war

them has received the same attention as the cyborg so far. It is also Haraway's recent work that reveals a different understanding of technology and how it relates to the body. In fact, it is precisely the idea of *relation* itself that is crucially reframed by Haraway. Rather than denoting the processes through which supposedly preexistent phenomena are connected with one another, or even merged into one, the very idea of relation is reframed as reading-into-each-other. Haraway elaborates that this idea has far-reaching consequences for the understanding of technology and the body. Instead of denoting a thing or even an artifact, technology is reworked as a relational *process*. Such an understanding of technology not only resembles much of Latour's notion of technology as referring to processes of mutual mobilization and reconfiguration between humans and nonhuman entities, as I have elaborated in chapter two, but also eschews the idea that technology would be something external to the body, something that would only cumulatively extend the body. Drawing on Maurice Merleau-Ponty's and Don Ihde's phenomenological reading of the body, Haraway puts forward the idea that we are "bodies in technologies", emphasizing that,

"technologies are not mediations, something in between us and another bit of the world. Rather, technologies are organs, full partners, in what Merleau-Ponty called 'infoldings of the flesh.' I like the word *infolding* better than *interface* to suggest the dance of world-making encounters. What happens in the folds is what is important. Infoldings of the flesh *are* worldly embodiment." (Haraway 2008: 249)

Far from signifying something external to the body, technologies denote the ways human and nonhuman bodies and lives are entangled with each other, "the knots we call beings" (Haraway 2008: 250). Haraway's figure of the material-semiotic actor, and in particular the concept of the apparatus of bodily production, demonstrates lucidly that a politics that draws on the idea of a supposedly natural body and on 'nature' as antithesis to technology and culture has to fail in its attempt at countering the practices of the contemporary global bio-genetic capitalism (see Braidotti 2013) that increasingly commodifies the generative forces and flows of organisms on a molecular level. Such a politics has to fail, precisely, because it keeps on telling the same old stories in which nature and material bodies are only passive objects, constantly threatened by technology and science, which in turn are understood as deeply traversed by, or even identical with, pure instrumental reason. This critique applies equally to conservative

work once again. Perhaps the same paper needs to be written again and again." (Haraway 2004: 5)

cultural pessimist approaches and to critical approaches standing in a Marxist tradition or in the tradition of the Frankfurt School. If, for example, Langdon Winner laments in his essay entitled *Resistance is Futile: The Posthuman Condition and its Advocates*[6] that natural bodies would become more and more technologized, he exactly buys into this fallacy.

In his essay, Winner differentiates between three perspectives on the human condition and technology: first the idea that humans are tool-making animals, second the claim that technologies are extensions of human organs (the telephone, for example, extends the ability to hear and speak, glasses extend the ability to see, and so on), and third the idea that human bodies and technologies have become inextricably entangled with one another to such a degree that the boundaries between nature and culture, organisms and technology, natural and artificial would lose much of their meaning. While Winner agrees with the first two perspectives as "common sense" (Winner 2005: 394), he strongly opposes the third one which he not only regards as counterintuitive but also as suggesting "that there is actually no meaningful boundary between humans and technology at all" (ibid) anymore. Particularly, in Haraway's figure of the cyborg and the idea that bodily boundaries and faculties are not determined by nature, Winner identifies a dangerous philosophical position that would not only play into the hands of global biocapitalism, but also grant the biotechnology industry full access to the body.[7] For Winner, such an approach "meshes nicely with the work of

---

6   The title of Winner's article clearly leans on a famous quote of the Borg, a hostile cybernetic species in the *Star Trek* universe, who introduce themselves at every encounter as follows: "We are the Borg. Lower your shields and surrender your ships. We will add your biological and technological distinctiveness to our own. Your culture will adapt to service us. Resistance is futile". The Borg are a collective of bio- and nanotechnologically altered assimilated species threatening large parts of the galaxy. Every assimilated entity was once a humanoid being that has been transformed against their will into a cybernetic organism functioning now as a drone of the Borg collective. The ultimate goal of the Borg is to transfer, meaning, to assimilate, every higher life form in the galaxy into an alleged ultimate state of perfection. Winner is clearly implying here that modern technology would threaten the very nature of our bodies and ourselves by transforming our bodies into genetically and technologically altered, emotion- and mindless cybernetic organisms without any individuality or personality.

7   Not only that everyone who has actually read Donna Haraway's work knows that this accusation cannot be further away from the truth. Haraway herself has opposed transhumanism and even problematized posthumanism as a not always helpful idea. For Haraway, posthumanism "is much too easily appropriated by the blissed-out, 'Let's all

radical reconstruction and recapitalization at stake in today's technical and corporate realms" (Winner 2005: 402) Going even further, Winner accuses scholars in science and technology studies not only of "enjoying" placing themselves "elbow to elbow with the scientific researchers and business leaders who move and shake with initiatives in globalization", but also of "clearly' siding "with the artificial" (ibid) instead of 'the natural'.

"For you see, dear friends, the boundaries have already been breached, the precedents established, the work of innovation set in motion, and the 'promising monsters' all introduced at the cyborgfeminist/science studies debutante ball." (Winner 2005: 408)[8]

---

    be posthumanists and find out next teleological evolutionary stage in some kind of transhumanist techno-enhancement.' Posthumanism is too easily appropriated to those kinds of projects for my taste. Lots of people doing posthumanist thinking, though, don't do it that way. The reason I go to companion species is to get away from posthumanism." (Gane/Haraway 2006: 140). Even more convincing however is the fact that Haraway's feminist engagements with science and technology were always critical to the biotechnology industry and technobiocapitalism. Take for instance her careful analysis of the role of the company DuPont in creating and distributing the OncoMouse™ in *Modest Witness*. It is in the same book where Haraway also states that, "The capacity for multisided, democratic criticism and vision that fundamentally shape the way science is done hardly seems to be on the political agenda in the United States, much less in the R&D budget of universities, in-house government labs, or industries—even while *how*, in fact, science is done is being reshaped in revolutionary ways." (Haraway 1997: 94). Just a few sentences later, Haraway stresses that, "Technoscientific democracy does not necessarily mean an antimarket politics, and certainly not an antiscience politics. But such democracy does require a *critical* science politics at the national, as well as at many other kinds of local, level. 'Critical' means evaluative, public, multifactor, multiagenda, oriented to equality and heterogeneous well-being. Nostalgia for 'pure research' in mythical ivory tower is worse than ahistorical and ideological" (Haraway 1997: 95).

8 Winner, however, is by no means the only (ironically, often male) critic who accuses feminist and queer scholars of not knowing what the best for 'women' would be. Hans Harbers (2003: 431), for example, even identifies "a remarkable coalition between feminist theory and medical technological practice: the latter almost appears to be the material realization of the former". For Harbers, feminist and queer scholars, instead of "deploying the individuality of the woman against the gradual growth of medical-technical invasions", would run "the political risk of not noticing the disappearance of women as autonomous individuals and their increasingly prominent presence as

Winner intensifies this critique to the point of implying that theorists of the posthuman condition such as Donna Haraway and Bruno Latour would seemingly not care much about the fact that, in this moment, more than a billion people are living in poverty, and do not have access to adequate medical assistance and clean drinking water. Expanding on this accusation, Winner argues,

"For anyone who cares to examine them [the conditions of humans living on Earth at present, JB], the data are chilling. According to the 2001 edition of the UN *Human Development Report*, 1.2 billion people on the planet suffer in extreme poverty, surviving on less than $1 a day, while a total of 2.8 billion (roughly half of the world's population) live on less than $2 a day. Some 2.4 billion people are without access to basic sanitation. Of the world's children, 325 million are out of school at the primary and secondary levels. For children under the age of five, 11 million die annually from preventable causes. Perhaps those now enthralled with cyborgs, hybrids, extropians, and posthumans will find such information insufficiently novel or thrilling to deflect their ambitious philosophical and research agendas. But the rest of us should take notice." (Winner 2005: 407)

Winner's message is clear: While people are dying from preventable causes, posthumanist and postmodern scholars would only care about pushing their own career with their sophisticated theories, merely exercising mental acrobatics and remaining blind in the face of social wrongs. Thus, for Winner it is obvious that "most of the benefit from such discourse appears to be career development for well-heeled intellectuals in Paris, Santa Cruz, Cambridge, and other R & D hubs" (ibid), ignoring that it was precisely Donna Haraway who argued that "research priorities and systems of research must be shaped from the start by people and priorities from many areas of social practice" (Haraway 1997: 93). More importantly, Haraway emphasized the necessity for "democratic criticism" and for developing "a stronger technoscientific democracy" (ibid: 115).[9]

'prosthetic bodies,' or even worse, it may legitimize it" (Harbers 2003: 431). While Harbers conlcudes that the question has to be how to avoid "this kind of political hazard of postmodernism" (ibid), feminist technoscience studies scholars such Donna Haraway are already two steps ahead with their emphasis on the necessity of situated, embodied accounts.

9   Haraway makes this point even more clear in an interview, saying: "A lot of my heart lies in old-fashioned science for the people, and thus in the belief that these Enlightenment modes of knowledge haven ben radically liberating; that they give accounts of the world that can check arbitrary power; that these accounts of the world ought to be in the service of checking the arbitrary." (Haraway 1991b: 2)

Not only that Winner fails to see that Haraway's cyborg figuration, precisely, does not stand for the celebration of genetic engineering or the technological and biomedical enhancement of the body, but rather for the deconstruction of stories about origins and purity which still haunt the bodies of everyone who has been marked as the Other. What is more, Winner even confuses cybernetics, posthumanism, and transhumanism; meaning, the science of self-organizing processes of information and signification in organic and mechanical systems, critical theories and tropes questioning the figure of the human, and technophilic fantasies of enhancing (certain) human beings. In an important sense, *post*humanism (at least in its critical and feminist tradition) does not mean *after* humanism in a temporal sense but rather going-beyond-with Enlightenment humanisms. As a philosophical position which roots in postmodern philosophy (in particular, feminist and queer approaches), posthumanist theories do not focus on technology itself (whatever that might be) but rather on dismantling traditional hierarchical dichotomies and boundaries such as human/nonhuman, mind/matter, organism/machine, active subject/passive object, self/other, that function as foundational assumptions of Enlightenment humanisms. In fact, it is for this reason that Enlightenment humanisms failed to fulfill their emancipatory and liberatory promises for all those who were considered merely as a reference point for the construction of the white, rational, bourgeois men: women, people of color, and animals.[10]

Posthumanist theories, thus, attempt to decenter and deconstruct what might be called traditional Western thought and discourse – that is, anthropocentrism, human exceptionalism, and what Donna Haraway (1991a: 189) has termed the 'god-trick' of seeing everything from nowhere. Transhumanism on the other hand denotes a movement that seeks to take control over human evolution through the means of technologies such as genetic engineering, mind uploading, cryonics, and so forth. As of today, the majority of these technologies are largely hypothetical and abstract ideas. This however does not and should not mean that they are not to be criticized and problematized. Instead of granting these images and imaginations more reality and subsequently more power by ignoring that at the moment they are precisely that, namely images and imaginations, it seems far more promising to highlight that exactly because this is one version of a future yet to come, it can also be prevented or rewritten. In any case, transhumanism can be regarded as a kind of technophilic hyper-humanism,[11] condensing and reifying rather than leaving behind traditional dichotomies. And indeed, Nick

---

10 Cf. also Braidotti (2013).
11 Cf. also Singer (2015).

Bostrom, a central figure of the transhumanist movement, states that transhumanism "can be viewed as an extension of humanism, from which it is partially derived. Humanists believe that humans matter, that individuals matter. [...] Transhumanists agree with this but also emphasize what we have the potential to become" (Bostrom 2003: 4). Who this 'we' is, however, remains largely open. As a highly corporatized faction sponsored amongst others by the biotechnology and pharmaceutical industry, the transhumanist movement markets itself under the label "humanity+" as an organization that advocates the use of a wide range of existing and future technologies (from drugs that enhance cognition and prenatal screening to such enigmatic sounding things as mind uploading) in order to expand human bodily capacities. Despite the fact that Bostrom and other transhumanist theorists are convinced that it is a "typical pattern with new technologies [...] that they become cheaper as time goes by", believing that not only those who can afford these technologies but eventually "everybody" would benefit (Bostrom 2003: 20), the concern that these (mostly hypothetical) technologies might establish a divide between bio-technologically augmented human beings and those who are not in the near future, cannot be swept away just like that. Apart from the fact that the picture of the future many transhumanist theorists draw is not only an overwhelmingly optimistic one but also one that ignores to a certain extent fundamental socio-economic dynamics, transhumanist efforts are clearly less democratic when it comes to the question whose lives are worth to become improved. Such positions, therefore, not only remain anthropocentric and human exceptionalist at their heart but also blindly perpetuate, if not indifferently reinforce, social inequality.

Ironically, Winner not only seems to share the same belief in the omnipotence of science and technology that underlies transhumanist accounts, but also a similar belief in technological determinism. Both, Winner and Bostrom, trust in the power of new technologies and technoscientific practices to rework our bodies and lives with far reaching consequences. However, while Bostrom can hardly wait for this future to come, Winner is certain that this future has to be prevented by all means necessary. What is more, it is also a very similar idea of humanism that Bostrom and Winner seem to share; and Winner has even sworn to defend against its posthumanist critiques. Even though transhumanism wants to overcome the limits of the supposedly natural human body, when it comes to the idea of the individual and the notion of the autonomous subject, transhumanism does not leave anthropocentric subject theories behind. On the contrary, transhumanism not only can be understood as a hyper-humanism but also as a hyper individualism.

In his call to abandon the terms cyborg, hybrids, and quasi-objects altogether and instead return to 'the human' and to the question of "what it means to be human in the first place" (Winner 2005: 405), Winner seems to follow Jürgen Habermas' later works. For Habermas human nature is fundamentally threatened by new biomedical technologies and the technosciences. It is not only "the dividing line between the nature we are and the organic equipment we give ourselves" (Habermas 2003: 22) that becomes increasingly blurred, but also the fact that certain new technologies could have the power to call into question the "physical basis which 'we are by nature'" (ibid: 28) that worries Habermas. To put it bluntly, what is at stake for Habermas is the choice between the nature we are by chance and the nature we give ourselves through the means of science and technology. In May 2012, Habermas was awarded with the *Erwin Chargaff Award for Ethics and Science in Dialog* in Vienna. In his address, he strengthened his position by stressing that genetics and new biotechnologies would lead to a limitless manipulation of the (human) body, eventually detaching the evolution of human nature from cultural learning processes. It is precisely against this background that Ian Hacking accuses Habermas of being a 'bio-conservative'. Habermas, however, never made a secret of the fact that he saw this label as quite fitting for him and his philosophical project. In a letter to Hacking from August 2009, Habermas even wrote that he "never thought that any version of 'conservativism' would apply to me, but 'bioconservativism' is a wonderful term!" (Hacking 2009: 14).

There are at least two problematic assumptions hiding behind Habermas's position. For one thing, Habermas's fear that the evolution of human nature could become independent from what he calls cultural learning processes implies that biological and cultural evolution can be separated neatly. Taking epigenetic research into account, however, this seems to be very unlikely. What is more, Habermas's argument implies that while biological evolution has to be regarded as a fate or as a gift (even where for countless people, in fact, it means to suffer a lifetime), cultural evolution represents the outcome of hard work, and therefore should not be up for purchase. For Habermas, the intrinsic worth of effort itself seems to be at stake with new biotechnologies and medical technologies. One might call this position a kind of biotechnological Calvinism.[12] For the other

---

12 In a similar way, also the philosopher of technology, Gernot Böhme, is convinced that constraints would not come any more from 'within', representing the fruits of hard work, meaning, self-disciplining, but from the 'exterior', namely from technologies: "The new constraints were constraints imposed by technical mechanisms of social existence, by technological infrastructure. […] in today's world, punctuality is no longer

thing, like many other scholars who see the need for defending the supposed naturalness of the human body or human nature itself, Habermas leaves largely open what, in fact, 'human nature' is that has to be defended against its alleged abolition today. As a matter of fact, human nature can mean very different things. It can, for example, stay for the biologically distinguishing characteristics which humans as a species have, such as the fact that the average body temperature lies between 36.5 and 37.4 degrees Celsius. Human nature can also stay for a supposed natural source for normative moral behavior, such as the claim that heterosexuality is 'natural' while homosexuality would be 'unnatural'. Finally, the term can also refer to a metaphysical assumption about the supposed essence of the human. However, while Habermas draws on Hannah Arendt's idea that being born is the precondition for social action[13] for characterizing his notion of 'human nature', arguing that as soon as human beings experience themselves as made, that is, as a product of someone else, rather than as being born by chance they would lose the capacity to 'start something new' and consequently their statues as an autonomous subject, it seems that for Langdon Winner 'human nature' is just 'human nature', something seemingly ahistorical, time-transcending, and essentially given; at least until modern technology and science interferes, taking in the function of a kind of counter-nature.

The idea of the technoscientific dissolution of the supposedly natural human[14] body represents one of the most pervasive tropes in critical theories on

---

    the hard-won fruit of systematic inner discipline, but an effect of external compulsion – a kind of by-product of the automatic action of the very technical conditions of daily life" (Böhme 2012: 27).

13 For Hannah Arendt it is clear that "[w]ithout action, without the capacity to start something new and thus articulate the new beginning that comes into the world with the birth of each human being, the life of man, spent between birth and death, would indeed be doomed beyond salvation" (Arendt 2000: 181).

14 Interestingly, the biotechnological and economic exploitation of the generative forces of the bodies of animals does not seem to bother many scholars in this critical tradition too much. While the term biocapitalism denotes a relatively new phenomenon closely associated with genetic engineering, it ignores that the commodification of the cognitive and material faculties of nonhuman bodies is already happening for a very long time. Take for instance dogs, cattle, and chicken that have been bred, that is, fundamentally adapted in their bodily shape, capacities, and behavior, to human needs. The bodies and lives of animals, thus, can be seen as biocapital par excellence since surplus value here is produced by means of the commodification of *bios*. In fact, the etymological root of the term capital itself refers to the trade and ownership of

technology and the body. Similar to Langdon Winner (2002: 106), who argues that technology "will eventually claim the human body itself" and "dissolve" it, also the critical theorist of technology Gernot Böhme (2012) warns us from an "invasive technification" and subsequently an alleged "dissolution" of the supposedly natural human body. For Böhme, as much as for Winner, this threat comes equally from the military-industrial complex as well as from postmodern feminist and other critical theorists. Believing that new technologies would "penetrate into the depths of who we are, into our bodies and our communicative relations", Böhme rhetorically asks if this is "what we want" (Böhme 2012: 7-8). Fundamentally misunderstanding and reducing the work of both Latour and Haraway to a mere caricature, Böhme replies to his own question,

"To be sure, there are people who greet the way that invasive technification is transforming human relations with euphoria. They have found their spokesman in Bruno Latour, a figure who glorifies the increasing obsolescence of the distinction between nature and technology, and in Donna Haraway, who empathically greets cyborgs as a new form of human life." (Böhme 2012: 8)

Leaving aside who this 'we' is, Böhme calls out for an "organized resistance" against "artificiality", turning to "natural values" instead (Böhme 2012: 161). What Böhme has in mind here is nothing less than the call to take nature "as a point of orientation" in order to be able again to "insist on 'the natural'" as a "valid basis for resisting total technification of human life and relations" (ibid: 9). However, even Böhme has to admit that there is no such thing as a 'natural nature', that is, a nature untouched by culture; at least not any more. Thus, if Böhme tries to recover the normative dimension of the concept of 'nature' in order to be able to defend "the nature we ourselves are: the body" against modern science and technology, he implies that technology and the body would be two completely different spheres which only recently have been 'molten' together, and that the body once was a completely 'untechnologized body'. And indeed, Böhme argues that,

"Until relatively recently in history, everything that grew out of it [the body, JB], was accepted almost as a brute fact [...] One was born as a man or a woman, with a particular constitution and particular dispositions, perhaps with certain illnesses as well. [...] Body,

---

animals. *Capitalis* in Latin originally meant 'of the head' and denoted the unit in which wealth was measured: namely the heads of cattle.

bodily constitution, illness, children – these were what primarily if not exclusively constituted one's fate." (Böhme 2012: 178)

Böhme's argument that until relatively recently in history one was born with a particular constitution such as, for example, a predisposition for a specific illness and had to cope with this fact as a kind of fate brings with it at least two flaws. First of all, it ignores that also several hundred years ago people were endowed with agency and, for example, serious illness was never a mere fate one simply had to accept. A quick look into the history of medicine demonstrates that the desire to overcome illness and death can be dated back at least as early as to the use of herbal medicine in the Stone Age or the origins of Indian Ayurveda 4.000 BCE. Second, it implies the morally and politically highly problematic idea that, even with the proper technologies today, one should not intervene in the biological constitution of the 'natural body'. Böhme demonstrates that he is serious about this last point by implicitly accusing Jean-Luc Nancy, who underwent a heart transplant, of masking his "personal", "or rather philosophical failure" (Böhme 2012: 231) – which would manifest in not accepting the fact of dying but instead letting transplant a foreign heart inside his body – by theorizing 'foreignness' and identity departing from the experience of his own heart transplant. The reason for this lies in the fact that for Böhme having a 'natural body' means "living in accordance with nature as an essential facet of one's own being […] paying due heed to something foundational in oneself that cannot be technically produced, reproduced or manipulated" (Böhme 2012: 163). But it is precisely through modern science and technology that this supposed naturalness of the body, and hence of human nature, would be threatened. How would it be possible to acknowledge that "I am my body", Böhme asks, "if all bodily organs, including the heart, can in theory be replaced" (ibid: 179)? Where once "nature meant the factually given, the *données* of fate, etc.", science and technology today would allow to intervene within this process, "totally dissolving" human nature (ibid).[15]

In the end, what Böhme does is to fundamentally oppose modern technology and the material body. While the material body (or more precisely, its 'inside') remains the last bastion of a supposed naturalness that has to be defended at all

---

15 Böhme's talk of a dissolution of human nature resembles what C. S. Lewis already in the early 1940s framed as "the abolition of man"; the fear that one day humans may obtain full control over their biological evolution, throwing overboard supposedly universal values and in doing so stepping outside the moral order which once had been dictated by Nature.

costs, modern technology represent a mere means of domination and oppression. This opposition becomes particularly apparent when Böhme argues that his theory would be critical "by its very nature" insofar as it understands technology as something that stands "in a relation of tension to what it is a technification of" (Böhme 2012: 19). To put it simply: There is modern technology on the one hand side and that which it 'technologizes', transforms, or even dissolves on the other. With such a move, however, (modern) technology not only becomes something external to the human body but also something that acts upon our bodies, transforming us into a kind of artificial monster. While Gilbert Simondon (1958; see also chapter two) wrote almost a half century ago that the machine is "a stranger to us", however, a stranger in which what is human is materialized and therefore remains human, for Böhme technology becomes a kind of prison for the proper human, eventually dissolving it.

Strangely enough, the trope of the technological dissolution of nature, both 'external nature' and 'the nature we ourselves are', has been and still continuous to be at the heart of many approaches in the tradition of the Frankfurt School. However, reading carefully the work of Theodor W. Adorno, Max Horkheimer, Walter Benjamin, and Herbert Marcuse, what becomes visible is, in fact, a far more ambivalent perspective on technology. Indeed it is possible to find passages in the works of the authors mentioned above where modern technology is equated with instrumental reason and dominance. This is particularly true for the philosophy of Adorno and Horkheimer.[16] But then there are also passages in which technology is understood in a positive or at least in an ambivalent way. Adorno, for example, emphasizes in *Über Technik und Humanismus* (2003 [1953]) that technology and society cannot be separated from each other, that the whole development of technology is "determined by society". Hence, Adorno not only anticipates the idea of the social construction of technology, but in fact also implies that technology, as a social activity and product, is always part of the human.

Also for Walter Benjamin technology, or more precisely certain technologies, can even assume the shape of a revolutionary force. In *The Work of Art in the Age of Mechanical Reproduction* (2008 [1936]), Benjamin argues that it is only through the process of technological reproduction that the work of art gets separated from its ritual function. Through the process of its *technological* reproduction (first through lithography, later through photography and film, and eventually, even though Benjamin did not get to see it, through digital technolo-

---

16 Especially in *The Dialectic of Enlightenment* (1972 [1947]) Adorno and Horkheimer largely equate technology with the domination of nature.

gies) the work of art becomes fundamentally questioned in its uniqueness, losing what Benjamin terms its "aura", that is, its particular presence in space and time, its unique existence at the very place and time where the work of art resides.[17] Against many misreadings of his work, Benjamin, precisely, does not see this necessarily as a loss but, on the contrary, also as a chance for democratizing both the access to art and art itself. Instead of only being a privilege for an elite, art becomes an event in which practically everyone can participate. As a Jewish Marxist who was forced to flee Nazi Germany and later also the Franco regime in Spain, Benjamin nevertheless was also very well aware of the power of new technologies such as film and broadcast for manipulating and controlling the masses. It is also against this particular background that he ends his essay worrying about the aestheticization of politics through fascism and the politicization of art in communism, highlighting the role of technology in dictatorial regimes.

Marcuse's understanding of technology resonates with Benjamin's fears. In *Some Social Implications of Modern Technology* (2004 [1941]), Marcuse differentiates between technology and technics. While for Marcuse technics refers to "the technical apparatus of industry, transportation, communication", technology denotes "a social process", the organization form of modern societies (Marcuse 2004 [1941]: 41). Even tough Marcuse argues that modern technology brings with it what he terms 'technological rationality', or technical reason – that is, a specific form of rationality which reduces life itself on the aspect of its mere usefulness – it is important to understand that Marcuse wrote these lines against a specific historical backdrop: namely, the emergence of populist mass movements in Europe (particularly, the rise of fascism) and the terrors of the NS regime. According to Marcuse, these phenomena can only be understood sufficiently if the role and function of modern technology is taken into account. In an important sense, however, for Marcuse, it is not technology itself that spawns fascism and totalitarianism. Quite the contrary.

"The technological process itself furnishes no justification for such a collectivism. Technics hampers individual development only insofar as they are tied to a social apparatus which perpetuates scarcity, and this same apparatus has released forces which may shatter the special historical form in which technics is utilized." (Marcuse 2004 [1941]: 63)

---

17 In a particular sense, it could even be said that the loss of the aura denotes the breaking up of a specific fabric or even an entanglement of human and nonhuman forces that only exist at a particular time and space.

Non only that Marcuse concludes that "all programs of an anti-technological character, all propaganda for an anti-industrial revolution serve only those who regard human needs as a by-product of the utilization of technics", he goes even so far as to emphasize that precisely for this reason the "enemies of technics", such as the philosophers of the so-called simple life, or the propagandists of blood and soil, would "readily join forces with a terroristic technocracy" (Marcuse 2004 [1941]: 63). By fighting technology, instead of specific socio-political-technical apparatuses, also the democratizing potentials of technologies are thrown overboard, and with them "the potential instruments that could liberate" us (ibid).

Reading the works of the Frankfurt School not only a different, that is to say, a more ambivalent picture of technology emerges, but it becomes also understandable why Adorno, Horkheimer, Marcuse, and Benjamin had plausible reasons to be critical against contemporary technologies.[18] It is this situatedness of their thoughts and concepts what gets lost in many approaches that all too often

---

18 Nevertheless, a serious problem of the Frankfurt School can be identified in their picture and understanding of the natural sciences and the practices of scientists. Not only that the Frankfurt School, in particular, Max Horkheimer in *Traditionelle und kritische Theorie* (1937), failed to raise the question for the constitutive conditions of the natural sciences and in doing so implied that it seems that for the natural sciences the specific social and economic contexts within which scientific knowledge and theories are produced would have no effect on the knowledge and the theories produced (Cf. also Becker/Jahn 2003). The Frankfurt School also believed that it would be impossible to produce and pursue critical theories – that is, theories that not only understand theory as a moment of social practice but also critically reflect the standpoint and situatedness of the scientist involved and explicitly take into account questions of inequalities, power, and dominance – within the natural sciences, precisely, because science in general has been one-dimensionally equated with positivism and instrumental reason. Consequently, the dividing line between the natural sciences on one side and the social sciences and the humanities on the other is simultaneously seen as the dividing line between traditional and critical theories (or at least as the possibility for the latter since for Horkheimer and Adorno many approaches and theories in the social sciences were, in fact, traditional bourgeois theories, and thus also 'traditional' rather than 'critical'). The work of many feminist, postcolonial, and other critical scholars trained in the natural sciences such as Evelyn Fox Keller, Ruth Bleier, Ruth Hubbard, Donna Haraway, Lynda Birke, Karen Barad, Myra Hird, Astrid Schrader, and Banu Subramaniam, to name but a few, however demonstrates that this story could also be told differently.

draw relatively ahistorically on the work of the Frankfurt School, trying to preserve and adapt it for the critique of, say, genomics or biotechnologies today. Donna Haraway speaks against this particular backdrop of,

"the stupidity of critical theories in just doing critique once again, in being stuck where Adorno and Horkheimer were much more legitimately stuck. What they did then needed to be done. But it is crazy to be stuck in that relentless complaint about technology and techno-culture and not getting the extraordinary liveliness that is also about us." (Gane/Haraway 2006: 141-142)

Haraway's argument might sound too harsh at the first glance but it certainly does not miss the point. By not only trying to preserve but also even to radicalize certain assumptions of the Frankfurt School in a rather ahistorical way, such positions run the risk of becoming themselves problematic, if not reactionary, today. If Kathrin Braun (1998), for example, argues in the tradition of the Frankfurt School that it irritates her that feminist and other critical scholars are "celebrating" the implosion of categorial boundaries, in particular, the boundary between nature and culture as well as between animal and human (Braun 1998: 167), once again the category of the human becomes completely ahistorical. While Haraway argues against a specific modernist mindset, Braun believes that Haraway would *factually* support attempts at not only "dissolving the category of the human" (ibid; trans. JB) but – by allegedly equating humans and animals – would also suggest that humans shall be treated like, for example, laboratory animals (ibid: 165). Similarly, also Regina Becker-Schmidt argues that, by deconstructing what it means to be human, Haraway's "anti-essentialist account" would, in the end, bring with it "the very same consequences" as "Utilitarian bioethics" (Becker-Schmidt 2000: 160; trans. JB), implying the opening up of an avenue for a kind of 'new eugenics'.

While from Heraclitus and Lucretius to Whitehead and Bergson to Deleuze and Haraway, 'nature' has been understood as a stream, as something that is constantly in motion, as a constant change or transition, the postulate of a return to human nature as well as to an allegedly technologically untouched natural human body, as it can be found in the work of Winner, Böhme, and others, represent the exact opposite: nature becomes the synonym for fixedness, ahistoricity, passivity, and consequently something that is constantly threatened by modern science and technology. What it means to be human, however, is not essentially given and determined once and for all but actively constituted together with nonhuman entities and forces; and thus through such diverse things as bacteria, environmental influences, political and religious ideologies, technologies, cultural norms, economic interests, and much more. What constitutes the specific human, therefore, is not a given essence or substance but the result of continuous struggles and negotiations in which not only humans take part In fact, it could even be said that 'we' only become human by constantly entering into relationships with nonhuman entities and things. This time, however, nonhuman entities are not merely a negative foil or antithesis for the self-creation of the human, as it is the case with traditional humanism. Once again, it is Haraway who reminds us that it is impossible to "do 'human' ahistorically [...], or as if 'human' were one thing. 'Human' requires an extraordinary congeries of partners. Humans, wherever you track them, are products of situated relationalities with organisms, tools, much else" (Gane/Haraway 2006: 146).[19] In a similar way also Latour states that "without elephants, plants, lions, cereals, ozone or plankton", 'we' would be a "[l]ess than a human. Certainly not a human" (Latour 1998: 235).

Therefore, invoking ideas of a natural state of the human seems not only epistemologically limiting but also ontologically reactionary in perpetuating a quasi-essentialist and naturalized understanding of what constitutes the human as well as human bodies and how they ought to be. What is more, once again the dichotomy of active mind and passive matter is brought into position here. The body remains mute and passive, purely ahistorical matter, while technology is understood as condensed social power relations or even as materialized instrumental reason. Consequently, such a rhetoric that appeals to the natural and the pure, to a body untouched by science and technology, cannot and "will not help—emotionally, intellectually, morally, or politically" (Haraway 1997: 62) because it only perpetuates the same old stories of naturalness, purity, and dominance which consumed so many lives in the past.

In *Specters of Marx*, Derrida asks what it means "to follow a ghost? And what if this came down to being followed by it, always, persecuted perhaps by the very chase we are leading?" (Derrida 2012: 9-10) What if the approaches on technology and the body problematized here and in chapter one are indeed following and followed themselves by a ghost? What if the invoked images of the body and how it ought to belong to a different time and space where the world and everything in it (as if the world would be a container) was made intelligible through a dualist mindset within which only one side could be the active one

---

19 Judith Butler makes a similar point in *Undoing Gender*, arguing: "For the human to be human, it must relate to what is nonhuman, to what is outside itself but continuous with itself by virtue of an interimplication of life. This relation to what is not itself constitutes the human being in its livingness, so that the human exceeds its boundary in the very effort to establish them (Butler 2004: 12).

while the other remained mute and passive forever? What if, in fact, material bodies "do not pre-exist as such" but rather materialize together with their particular boundaries always only through "mapping practices", as Haraway (1991a: 201) suggests? And what if technology, far from being something external to the body, is always, yet in different ways and with very different consequences, entangled with material bodies? Is it possible to arrive at a deeper understanding of the processes through which bodies come to matter in both senses of the word in their entanglement with particular technologies and technoscientific practices, from which in turn the bodies concerned cannot be separated? Does the concept of the apparatus of bodily production allow for such an understanding of technologies and bodies as always already entangled with one another; and with it for an understanding of technology as something 'foreign' and yet always also a part of us?

Even though Haraway brings to the fore that "the world encountered in knowledge projects is an active entity" (Haraway 1991a: 198), a "witty agent",[20] the figure of the apparatus of bodily production itself remains conceptually and methodologically vague. Building on Haraway's insights, the feminist quantum physicist Karen Barad further elaborates the figure of the apparatus of bodily production, highlighting the inseparability of knowing and being. Reading diffractively[21] insights and concepts from quantum field theory, feminist epistemology, and poststructuralist philosophy, as I will outline in what follows, Barad brings to the fore the ethical and political dimensions of processes of mattering.

APPARATUSES AS BOUNDARY-DRAWING PRACTICES

In the midst of the Second World War, Niels Bohr and Werner Heisenberg, two of the most influential quantum physicists of the 20th century, met in Copenhagen to discuss the efforts of Nazi Germany in building the atomic bomb. Bohr was not only of Jewish ancestry but also involved in the Danish resistance against the Nazi occupying forces. Heisenberg, on the other hand, was working on the "military applications of atomic energy" (Bohr 2005: 13). Worried about

---

20 It is important to understand here that Haraway's project was never a mere epistemological one; even though her work, in particular *Situated Knowledges*, has been often read in such a reductionist way. In fact, Haraway's situated epistemology cannot be separated from a relational process ontology, and vice versa. Likewise, epistemology and ontology, for Haraway, cannot be separated from question of ethics and politics.

21 For 'diffraction' as methodology see Barad (2007) and van der Tuin (2011; 2014).

the possibility that Nazi Germany might "utilize nuclear energy as a weapon of war" (ibid: 244), that is, build the atomic bomb,[22] Bohr cut all connections to Heisenberg before he left Denmark in October 1943, after he had been arrested as an enemy of the Nazi regime for helping refugees to leave the country and hence was facing the immanent deportation to the concentration camps. Bohr and Heisenberg's relationship would never recover from this meeting.[23] What remained influential well beyond the field of quantum physics until today, however, is the Copenhagen interpretation, the attempt of Bohr and Heisenberg to find an answer to what became known in quantum mechanics as the wave-particle duality.

Particles are entities that occupy a specific location in space and time. Waves on the other hand occupy more than one position at any moment in time. Waves therefore not only extend in space and time but also overlap or interfere with one another. In 1924, Louis de Broglie, a French physicist, suggested that not only photons – that is, small particles of electromagnetic energy with no mass and no electric charge whatsoever – but all matter has not only particle but also wave properties. The wave-particle duality, thus, indicates that every quantic entity (photons, electrons, and even atoms) exhibit the properties of not only particles but also waves (see de Broglie 1939); a statement that runs counter to Newtonian physics where, for example, light consists of particles but has no wave properties.

---

22 Heisenberg stated that he was neither sympathizing with the Nazi regime nor was he working on the atomic bomb, but only interested in "rebuilding science" after the displacement and deportation of Jewish scientists in Germany and Austria. As a matter of fact, Heisenberg was indeed not only observed and questioned but also defamed as a "white Jew" and a "representative of the Einsteinian 'spirit'" (Cassidy 2009: 269) in *Das Schwarze Korps*, the official newspaper of the SS, during the time of the Nazi dictatorship. On the other hand, not only Heinrich Himmler was convinced that Heisenberg would be loyal to Nazism, meaning, that he would be useful in providing the regime with the "practical benefits" of quantum mechanics "for the war effort" (ibid: 331), but Heisenberg himself has accepted the position of scientific director of the German nuclear project (the so-called *Uranverein*) in late 1942.

23 A number of letters to Heisenberg that Bohr wrote between the late 1950s and the early 1960s, but never sent, demonstrate that Bohr did not take it easy to end a personal friendship that lasted for decades. For the draft letters see the website of the *Niels Bohr Archive* (http://www.nba.nbi.dk/papers/docs/cover.html) accessed June 20, 2018.

Both Heisenberg and Bohr developed their own theories on this phenomenon. Heisenberg argued in his famous uncertainty or indeterminacy principle that the more precisely the position of an electron is known, the less precisely will its momentum be known, and vice versa. According to the uncertainty principle, it is therefore not possible to determine the exact position *and* momentum of an electron *at the same time*. What is more, Heisenberg believed that the very process of measurement would always and necessarily influence the result of measurement.[24] Bohr too was convinced that position and momentum are complementary properties and therefore cannot be measured at the same time. By means of his famous double-slit thought experiment,[25] however, Bohr argued that if electrons are directed through an apparatus which consists of two platforms behind one another (the first one having a single slit, the second one having two slits), the electrons will behave in one case like waves and generate a so-called diffraction pattern on the photographic plate behind the two platforms, and in the other case like particles, by generating a pattern characteristic for particles. Which pattern is generated depends on the measuring apparatus. It is, however, not possible to have it both: that is, to know which slit an electron went through *and* to know the pattern it produces on the photographic plate at the same time. In contrast to Heisenberg, however, Bohr concluded that it is not the case that measurements would *influence* the objects of observation but rather, that it is, in fact, only through the process of measurement – that is, through a particular constellation of the objects of observation and the agencies of observation – that a specific phenomenon (meaning, particular properties of the object measured) is *enacted*.[26]

It is precisely this idea that constitutes the point of departure for Karen Barad's attempt at developing a theory which, informed by quantum mechanics as well as poststructuralist and feminist philosophy, aims at leading "us out of the morass that takes absolutism and relativism to be the only two possibilities" (Barad 2007: 18), and in doing so promises to provide us with an understanding of the entanglement of knowing and being, as well as of matter, politics, and ethics. In contrast to Heisenberg's uncertainty principle, which is an "epistemic principle" because "it says there is a limitation to what we can know" (Barad

---

24 For a detailed description of the uncertainty principle see the pages 44-58 in Heisenberg (1958).
25 It was not until the early 1960s that it became technologically possible to perform the double-slit experiment with electrons, and later also with atoms and even molecules.
26 For an introduction in Bohr's correspondence principle see, for example, Plotnitsky (2013) as well as chapter three in Barad (2007).

2007: 116), what Bohr's insights suggest in Barad's view is nothing less than the idea that there is no stable observation-independent reality with preexisting properties but that it is only through particular measuring apparatuses that a certain phenomenon with specific properties becomes determinate – and only at the cost of excluding an other phenomenon with its own, distinct properties. This idea boils down to the consequence that, according to Bohr, the observer is always part of what she observes and seeks to understand. What follows from this for Bohr, and consequently also for Barad, is "that there is no unambiguous way to differentiate between the object and the agencies of observation. As no inherent cut exists between object and agencies of observation, measured values cannot be attributed to observation-independent objects" (ibid: 196).

Rather than being just another iteration of relativism, Bohr's philosophy-physics[27] not only calls into question a particular well-established tradition of Western metaphysics, namely "the belief that the world is populated with individual things with their own independent sets of determinate properties" (Barad 2007: 19), but according to Barad also provides us with what might be called a "proto-performative account of scientific practices" (ibid: 195). As such, Bohr's approach would aptly demonstrate that knowledge production is a naturalcultural socio-material process through which not only knowledge about the world is produced but also specifically re(con)figured parts of what counts as real or reality are performatively enacted.

Performative approaches contest representationalism, that is, "the idea that representations and the objects (subjects, events, or states of affairs) they purport to represent are independent of one another" (Barad 2007: 28). "Representational thought is analogical", as Brian Massumi (2004: xi) highlights; it aims at establishing a correspondence between world and words, nature and culture, matter and discourse, bodies and technologies. As a consequence of the Cartesian

---

27 With the term philosophy-physics, Barad highlights that for Bohr philosophy and physics cannot be separated from one another (Barad 2007: 24). This however, does not mean that philosophy and physics would be essentially the same thing. Rather, what the term designates is the idea that "philosophy is integral to physics […] In other words, physics without philosophy can only be a meaningless exercise in the manipulation of symbols and things, much the same as philosophy without any understanding of the physical world can only be an exercise in making meaning about symbols and things that have no basis in the world" (ibid: 68). Trevor Pinch brings a similar argument, stating that what is especially fascinating about quantum physics and its history is that "it has switched from physics to philosophy and then back again to physics. And these days it is more than just physics" (Pinch 2011: 434).

division between the internal and the external, representationalism lies at the heart of both scientific realism and social constructionism. Both scientific realism and social constructionism rely on mediators to bridge the epistemological and ontological gap between words and world, subjects and objects, representations and that which they represent. Consequently, representationalism is one of the fundaments of what Bruno Latour calls the "modernist settlement".[28] Where scientific realism and social constructionism would only differ is "on the question of referent, whether scientific knowledge represents things in the world as they really are (i.e., nature) or objects that are the product of social activities (i.e., culture)", as Barad (2007: 48) argues. Performative approaches circumvent the need for such a correspondence between world and words, matter and discourse by focusing on the question of how not only meanings but also particularly re(con)figured bodies, identities, and hence realities, are enacted through particular generative practices.

The concept of performativity can be traced back at least to Friedrich Nietzsche who problematized what he termed the "metaphysics of substance". For Nietzsche, no "substratum exists; there is no 'being' behind doing, effecting, becoming; 'the doer' is merely a fiction added on the deed—the deed is everything" (Nietzsche 1989 [1887]: 45). Since for Nietzsche doing preexists being, the latter would be nothing more (but also nothing less!) than congealed doing. Drawing on the work of Nietzsche and John Austin, who coined the concept of performativity for speech acts that enact that to which they refer,[29] Judith Butler demonstrates how identities, in particular gender identities, but also bodily materialities, are a matter of the repetition of performances. There is no essence behind doing or becoming. Each doing is a repetition, a reiteration, producing particular identities and their corresponding bodies. For Butler, gender and the gendered body is a doing, a becoming, rather than a preexisting, supposedly natural and thus ahistorical fact. Butler's notion of performativity is closely linked to her concept of materialization. Materialization, for Butler, denotes the inextricable, constitutive entanglement of matter and discourse. Matter is always materialized matter. That is to say, it is only through particular processes of materialization that matter "stabilizes over time to produce the effect of boundary, fixity, and surface" (Butler 1993: 9). Butler makes this idea explicit by referring to the medical interpellation of an infant at birth by a physician or midwife. By calling out

---

28 See Latour (1993; 1999a). See also chapter two for a detailed discussion of Latour's critique of what he terms the "modernist settlement".

29 Beside Nietzsche and Austin, also the work of John Searle, Jacques Lacan, and Jacques Derrida has been of crucial importance for Butler's notion of performativity.

"It's a girl!" or "It's a boy!" the infant shifts "from an 'it' to a 'she' or a 'he'" (ibid: 7). In contrast to a mere description or representation, discursive practices[30] enact that which they name by referring to cultural norms and systems of belief. Through interpellation social norms and identities (such as, for example, particular gender identities) are inscribed on the body of the individual concerned – even against their will.

Butler provides a powerful account of how discursive practices materialize what 'we' call reality. Discursive practices, however, should not be confused with mere linguistic statements. Importantly, Butler's performative understanding of discursive practices does not equal linguistic idealism. Neither does Butler reduce bodies or reality altogether to language, nor does she deny the materiality of the body – an accusation that Butler received predominantly often in the German speaking context. On the contrary, what Butler problematizes in *Bodies that Matter* (1993) is precisely the idea of a mere *social* construction of the body – that is, the belief that there is a natural body (we cannot know anything about) which than somehow becomes 'overwritten' by the social and cultural.

Even though Butler challenges social constructionist theorizations of the body with her notion of materialization, for Barad, she seems less successful in providing a convincing account of "the material constraints and exclusions, the material dimensions of agency, and the material dimensions of regulatory practices" (Barad 2007: 192). In doing so, Barad sees Butler's account running the risk of "reinstalling materiality in a passive role" (ibid).[31] Consequently, even though Barad regards Butler's account as a highly productive one, she also emphasizes that a serious difficulty arises from such a conception of performativity. Butler provides a highly productive account on how discourse comes to matter, that is, how discursive practices materialize what counts as real; but what about

---

30 In *The Archaeology of Knowledge* (1972) and later in his inaugural lecture entitled *L'ordre du discours* at the Collège de France, Foucault develops the term discursive practices to describe the very practices through which what counts as true and real in a particular place and time is considered as socio-culturally produced. Discursive practices, therefore, designate the practices that define and produce their referents.

31 Barad emphasizes that she, explicitly, wants to distinguish her critique "from a host of accusations against Butler that incorrectly accuse her of idealism, linguistic monism, or a neglect or even erasure of 'real flesh-and-blood bodies.' It would be a gross misunderstanding of Butler's work to accuse her of collapsing the complex issue of materiality to one of mere discourse, of arguing that bodies are formed from words, or of asserting that the only way to make the world a better place is through resignification" (Barad 2007: 192).

matter? "Butler correctly calls for the recognition of matter's historicity", Barad argues, but "ironically, she seems to assume that it is ultimately derived (yet again) from the agency of language or culture" (Barad 2007: 64). If matter is nothing else than the congealing or sedimentation of discursive practices, Butler's understanding of performativity and materialization suggests that there is no matter and material body prior to their enactment through particular discursive practices, or at least that we cannot know anything about it. If there is no matter and material body prior to its discursive production, consequently there also cannot be such a thing as a material or bodily agency, or at least no way to theorize it and to take it into account.

What is at stake is therefore the nature of matter and agency. Or, to put it another way, what Barad is interested in is the question of whether matter is solely the effect of human (and human alone) practices of knowing and being; or is there a way to theorize matter as dynamic and therefore as being involved in the processes of its own becoming. Is there a way to get to a different *materialist*[32] understanding of matter, yet without resurrecting the two zombies substantialism and essentialism? What is more, if what counts as 'true' and 'real' has to be regarded as the sole product of discursive practices what gets lost is the share of nonhuman entities in the processes through which not only (scientific) knowledge is produced but also that which counts as 'real' and 'reality' is continuously enacted. Similarly to scientific realist and social constructionist approaches also to theories of performativity, thus, applies that "the world is precisely what gets lost" (Haraway 1992: 313) when they ignore the share of nonhumans in materializing practices.

It is against this backdrop that Barad puts forward the necessity for a posthumanist account of performativity that also incorporates the nonhuman. As stated before, in Bohr's work Barad identifies a solid basis for such a posthumanist performativity that would allow for not only a deeper understanding of the epistemological but also of the ontological dimensions of technoscientific and other practices. Even though Barad contests the idea of an inherent, and therefore ahistorical, distinction between the microscopic realm and the macroscopic world, she is very careful in not drawing an analogy between "particles and people, the micro and the macro, the scientific and the social, nature and culture" (Barad 2007: 24). Neither does society mark a mere epiphenomenon that could be "explained in terms of collective behavior of massive ensembles of

---

32 Barad clearly situates her work in a materialist tradition and her thought "as being very much indebted to rich histories of materialist thinking" (Barad 2012a: 13), and in particular to the work of Marxist feminists.

individual entities (like little atoms each)" (ibid), nor is it the case for Barad that physics, biology, or chemistry could provide us with a sufficient account of the social, political, and ethical.[33] At the same time, Barad is very specific in emphasizing that it would be false to assume that quantum mechanics does only explain phenomena in the microscopic world, and therefore would be incommensurable with Newtonian physics, which in turn would only apply to macroscopic phenomena, say, to explain such things as the trajectories of rockets or ballistic missiles. As Barad explains, such a statement is not only false because Newtonian physics would be "a flawed theory" but nevertheless "a useful computational tool" – simply because it is much easier to apply than quantum mechanics and still provides in many cases "numeric values for some quantities that are excellent approximations to values calculated using the laws of quantum physics" (Barad 2007: 423, Fn. 22) – but also highly problematic because it reinforces a Cartesian metaphysics with the belief in an inherent distinction between the microscopic and the macroscopic world, along with the distinction between interiority and exteriority, matter and discourse, human and nonhuman. It is for this reason that I do not understand Barad's project as an attempt at imposing the theories and findings of quantum physics on science studies and the philosophy of science, as, for example, Trevor Pinch (2011) believes,[34] but as a kind of

---

33 In this context, Barad also problematizes the romanticization of quantum physics in popular literature, which would "too quickly [forget] that quantum physics underlies the workings of the A-bomb, that particle physics (which relies on quantum theory) is the ultimate manifestation of the tendency toward scientific reductionism, and that quantum theory in all its applications continuous to be the purview of a small group of primarily Western-trained males. It is not my intention to contribute to the romanticizing or mysticizing of quantum physics" (Barad 2007: 67-68).

34 Pinch acknowledges Barad's project as a "fascinating, complex, and important" theory, offering a novel way "of dealing with living in a material world" (Pinch 2011: 440). At the same time, he also argues that by "drawing upon the results of physics not as a metaphorical enterprise but as having direct implications for science studies", Barad runs the risk of courting "a form of scientism" (ibid). Pinch's greater problem, however, seems to be that, while Barad calls for a more situated account of scientific practices, and consequently of science she would "fail to situate the very part of science she is talking about, while drawing in a realist mode upon experiments to support her position" (ibid: 439). While it is true that Barad tends to be more euphoric (and sometimes less critical) toward theories and experiments in physics in contrast to, for example, sociological theories (which she often deconstructs thoroughly), I do not see her imposing the findings and theories of quantum physics on science studies or even

thought experiment itself which materializes reiteratively through a diffractive reading of quantum mechanics, feminist theory, and poststructuralist philosophy.

Being convinced that Bohr's concepts would not only challenge Newtonian physics with its idea that there is an observation-independent reality whose truths could be revealed through the proper scientific methods (that is, through specific instruments and experiments) but, in fact, also Cartesian epistemology *and* ontology with its "representationalist triadic structure of words, knowers, and things" (Barad 2007: 97), in her work, Barad reads Bohr's epistemological insights, poststructuralist philosophy, and feminist theory diffractively through one another, arriving at a posthumanist understanding of performativity. Instead of mirroring an ontologically preexisting reality, Barad's posthumanist understanding of performativity shifts the analytical focus to the material-discursive practices through that which counts as real is performatively enacted. The term 'posthumanist', here, signals in the wake of Donna Haraway's work on the material-semiotic actor that "nonhumans play an important role in naturalcultural practices, including everyday social practices, scientific practices, and practices that do not include humans" (Barad 2007: 32). In Barad's view, what Bohr rejects with his 'philosophy-physics' – and in particular with his argument that there is no measurement independent reality but, on the contrary, it is only through the practices of measurement that specific material re(con)figurings of what is seen as reality are enacted – is an "atomistic metaphysics that takes 'things' as ontologically basic entities" (Barad 2003: 813). Bohr would reject that the ontological structure of reality is build up by individual entities with clear boundaries and inherent properties. What follows from this for Barad is that a truly posthumanist understanding of performativity as well as a truly

---

on society. What is more, Barad rejects Pinch's argument that the practice of doing science and writing about science were "mutual exclusive" practices. After all, feminist scholars were the ones who demonstrated that it is possible to do science and to write about science at the same time. However, in contrast to sociological analyses of scientific practices many of these feminist scholars do not hold as their "first priority the proper description of what it is that scientists do", but instead raise the question: "How might science be practiced more responsibly, more justly? This issue is my passion, which is what drew me as a scientist into the discussion in the first place", Barad (2011: 450) states.

relational ontology has to challenge the metaphysical statement that relations require relata.[35]

To understand this, it is important to return once again to Bohr's double-slit thought experiment. As stated above, Bohr argued that position and momentum of a given photon can only be measured separately from each other: position can only be determined if it is measured with the help of an apparatus with a fixed platform, while in order to measure momentum an apparatus with a movable platform is required.[36] What follows from this is that for Bohr observation involves "an indeterminable discontinuous interaction", meaning, "there is no unambiguous way to differentiate between 'the object' and the 'agencies of observation.' No inherent/Cartesian subject-object distinction exists" (Barad 2007: 114). Therefore, it can neither be said what exactly constitutes the object of measurement, nor is it possible to determine the boundaries of an object of measurement. It is only through the introduction of a specific cut (an *agential cut*, rather than a Cartesian one, in Barad's terminology) that the object of measurement and the agencies of observation are separated from each another.

This idea has not only far-reaching epistemological but also ontological consequences; which is also the reason why Barad speaks of her approach as an "onto-epistemology", as the study of the inseparability of practices of knowing and being.[37] Instead of preexisting ontologically separable entities with inherent properties, *phenomena* mark the "primary ontological unit" (Barad 2007: 139) in Barad's philosophy. *Phenomena* are neither to be understood in a Kantian sense as the *appearance* of things-in-themselves (noumena), nor are they to be understood in a phenomenological sense as objects-as-they-appear, or noema (that is,

---

35 Influenced by the process philosophy of Alfred North Whitehead, Donna Haraway emphasizes a very similar idea, arguing that beings/bodies "do not preexist their relatings" (Haraway 2003b: 6).

36 See chapter three and chapter seven in Barad (2007) for a detailed description of Bohr's double-slit thought experiment and its physical and epistemological implications.

37 Understanding epistemology and ontology as entangled with one another does not mean to conflate them. Entanglements in an important sense, "do not erase differences" (Barad 2014: 176). The notion of entanglement, in a Baradian sense, is not about "the intertwining of two (or more) states/entities/events, but a calling into question of the very nature of two-ness, and ultimately of one-ness as well. [...] One is too few, two is too many" (ibid: 178). Consequently, in a similar way to wave and particle, for Barad, the relationship between the epistemological and the ontological has to be regarded as a relation of in/determinacy to one another.

the contents of thoughts), with the consequence of remaining faithful to the Cartesian dualism of subject and object, (absolute) interiority and (absolute) exteriority, as well as mind and matter. Rather, the term *phenomena* denotes the effects of the dynamic *intra-active* (not interactive) material-discursive relations; *"phenomena are the ontologically inseparability of agentially intra-acting 'components.'* That is, phenomena are ontologically primitive relations—relations without preexisting relata" (Barad 2003: 815). As such, *phenomena* are "material entanglements" (Barad 2007: 441, Fn. 13).

It is for this reason that in Barad's onto-epistemology it is only "through specific agential intra-actions that the boundaries and properties of the 'components' of phenomena become determinate and that particular embodied concepts become meaningful" (Barad 2003: 815). Because relations precede relata (which emerge in and through relations) in Barad's account, not only atomist metaphysics but also representationalism is turned upside down. What is more, in contrast to traditional theories of performativity, in Barad's posthumanist understanding, *phenomena* are not the sole product of discourse or language but always also the product of non-discursive practices. *Phenomena* are thus not only enacted by what used to be called the autonomous subject but also by what used to be regarded as mere passive objects – meaning, material bodies and their forces and flows, nonhuman entities, and even things.

Such a posthumanist understanding of performativity, with a reworked epistemology and ontology lying at its heart, promises in Barad's perspective the possibility for the taking into account of "the boundary-making practices by which the differential constitution of 'humans' and 'nonhumans' are enacted" (Barad 2003: 818). Barad very well knows that "putting forward an ontology" and "making metaphysical claims" is a risky undertaking, and that the "talk about the 'real' at the beginning of the 21st century may be the source of such a discomfort that it always needs to be toned down, softened by the requisite quotation marks" (Barad 2007: 205).[38] At the same time, she is also convinced that "'we' cannot afford not to talk about 'it'" (ibid); precisely, because it matters *how* reality is understood. In Barad's view, epistemology is not enough to understand the processes through which difference comes to matter, and marks are left on bodies along with their political and ethical consequences. What lies at the

---

38 It is for this reason that Barad's approach is neither a constructionist/relativist nor a realist one. Rather, with her deflationary account of knowledge, her situated notion of materiality and discursivity, and her posthumanist concept of performativity, Barad eludes and abrogates precisely the dichotomy of realism and constructionism itself, as Peter Wehling (2006: 96) points out.

core of mattering, as Barad explains, is not epistemological uncertainty but ontological indeterminacy, "a radical openness, an infinity of possibilities" (Barad 2012b: 16). Ontological indeterminacy does not mean that differences between epistemology and ontology, knowing and being, culture and nature, subject and object are collapsed or even conflated, but rather that they are mutually co-constitutive.

It is the concept of the *agential cut*, or *agential separability*, that allows for "a stronger *ontological* understanding of objectivity" (Barad 2007: 173). Agential separability designates "*an agentially enacted ontological separability within the phenomenon* [...] The crucial point is that the apparatus enacts an agential cut—a resolution of the ontological indeterminacy—within the phenomenon, and agential separability—the agentially enacted material condition of exteriority-within-phenomena—provides the condition for the possibility of objectivity" (Barad 2007: 175). In short, it is only through specific intra-actions that the boundaries between 'subject' and 'object', culture and nature, human and nonhuman, are enacted. In contrast to Cartesian cuts, these resolutions have to be understood as always local and temporary, meaning, as situated rather than as transcendent. The phenomena enacted are therefore inextricably bound up with the apparatuses through which they come into being. It is for this reason that in Barad's approach objectivity always entails responsibility and accountability for that which is brought into existence as well as for that which is rendered impossible in the very same moment.

Being responsible for the world and its becoming does not mean that the world is the product of 'our' choosing but rather that "reality is sedimented out of particular practices that we have a role in shaping and through which we are shaped", as Barad (2007: 390) emphasizes. Because responsibility, therefore, means intervening "in the worlds becoming" by reworking "what matters and what is excluded from mattering" (Barad 2003: 827), objectivity cannot be "a matter of seeing from somewhere, as opposed to the view from nowhere (objectivism) or everywhere (relativism)" (Barad 2007: 376),[39] but is precisely reworked as accountability for what comes to matter as well as for what does not. That being so, the objective referents in Barad's theory of agential realism are

---

39 It is precisely for this reason that situated knowledge, as Barad argues, "is not merely about knowing or seeing from somewhere (as in having a perspective) but about taking account of how the specific prosthetic embodiment of the technologically enhanced visualizing apparatus matters to practices of knowing" (Barad 2007: 470, Fn. 45) *and* being, one might add, since epistemological reflections have ontologizing effects.

not individual preexisting entities with inherent properties but *phenomena*, of which the corresponding apparatuses are inextricable parts. As mentioned earlier, apparatuses are always part of what they measure, as it is only through a given apparatus that particular properties and boundaries of what is being measured manifest (at the exclusion of others). Because measurements in Barad's account of technoscientific practices are "not simply revelatory but performative" (Barad 2012b: 6), they not only enact a particular knowledge about reality but also a particularly re(con)figured aspect of that reality.

In Barad's account, to theorize does not mean "to leave the material world behind and enter the domain of pure ideas where the lofty space of the mind makes objective reflection possible. *Theorizing, like experimenting, is a material practice*" (Barad 2007: 55). It is therefore not the world that has to adapt to theories about it but, on the contrary, theory always has to be adapted to the openness and the constant becoming of the world. Precisely because theories, rather than being "metaphysical pronouncements on the world from some presumed position of exteriority [...] are living and breathing reconfigurings of the world" (Barad 2011: 451), it would also make no sense to shield them from the world. This idea becomes especially apparent in Barad's understanding of the notion of the apparatus which shares a conceptual genealogy with not only Bohr's apparatus as well as Haraway's figure of the apparatus of bodily production, but also with Foucault's apparatus (or *dispositif*[40] in French).[41] Foucault uses the term apparatus in his analytic of power to describe the relationship between what is often regarded as two disparate phenomena: discursive practices and non-discursive ones. In "The Confessions of the Flesh", an interview from the year 1977, Foucault characterizes the apparatus as,

"a thoroughly heterogeneous ensemble consisting of discourses, institutions, architectural forms, regulatory decisions, laws, administrative measures, scientific statements, phil-

---

40 In a lecture from 2005, Giorgio Agamben argues that he is not satisfied with the English translation of 'dispositif', proposing a different "probably monstrous translation" which is "nearer to the French original", namely: "dispository" (Agamben 2005).

41 In another place, Barad states that her work combines "Bohr's notion of apparatuses as physical-conceptual devices that are productive (and part of) *phenomena* with Foucault's post-Althusserian notion of apparatuses as technologies of subjectivation through which power acts, and with Butler's theory of gender performativity which links subject formation as an iterative and contingent process to the materialization of sexed bodies" (Barad 2001: 86).

osophical, moral and philanthropic propositions—in short, the said as much as the unsaid. Such are the elements of the apparatus." (Foucault 1980: 194)

The apparatus, for Foucault, denotes "the system of relations that can be established between these elements [...] the nature of the connections that can exist between these heterogeneous elements" (Foucault 1980: 194). In what follows, Foucault distinguishes the *apparatus* from *episteme* (or discursive formations) by emphasizing that while the latter denotes a "specifically *discursive* apparatus" the former comprises of both the discursive and the non-discursive (ibid: 197). In an important sense however, it appears that for Foucault the non-discursive (and the material, in particular) would only somehow *support* the discursive – however, even the later Foucault remains unclear about the question of how exactly. In any case, the latter is clearly privileged above the former in Foucault's account.

It cannot be denied that Barad's understanding of apparatuses as *material-discursive practices* is clearly influenced by Foucault. As a matter of fact, Barad highly values Foucault's analytic of power, however, she also problematizes that,

"there are crucial features of power-knowledge practices that Foucault does not articulate, including the precise nature of the relationship between discursive practices and material phenomena; a dynamic and agential conception of materiality that takes account of the materialization of all bodies (nonhuman as well as human and that makes possible a genealogy of the practices through which these distinctions are made); and the ways in which contemporary technoscientific practices provide for much more intimate, pervasive, and profound reconfigurings of bodies, power, knowledge, and their linkage than anticipated by Foucault's notion of biopower (which might have been adequate to 18th practices, but not contemporary ones)." (Barad 2007: 200)

While Foucault falls short on his promise to provide an understanding of precisely *how* the discursive and the non-discursive are connected, Barad offers a more promising way to think both discursive practices and material phenomena as always already entangled with one another. In Barad's understanding, discursive practices and material phenomena, are not two distinct spheres; nor can they be reduced to one another. The notion 'material-discursive' rather signifies that "materiality is discursive (i.e., material phenomena are inseparable from the apparatuses of bodily production: matter emerges out of and includes as part of its being the ongoing reconfiguring of boundaries), just as discursive practices are always already material (i.e., they are ongoing material (re)configurings of the world)" (Barad 2003: 822). Material-discursivity, therefore, has to be understood similar to Haraway's notion of "the naturalcultural" as one word, as an inextricable entanglement; rather than the connection of two distinct phenomena. Reframing discursive practices and material phenomena as always already entangled with one another would make it impossible to privilege one above the other.

As mentioned before, Barad's notion of the apparatus shifts the focus to the question of how not only knowledge but also particular materialities with specific political and ethical consequences come to matter through particular material-discursive practices. In contrast to both Foucault and Butler, however, who focus solely on the materialization of human bodies, while they simply take for granted the constitution of nonhuman ones and the question how nonhuman bodies come to matter and become meaningful, Barad's account is not restricted to the domain of the social.[42] Rather, what Barad foregrounds are the complex intra-actions of human and nonhuman agencies in their entanglement with one another. In doing so, Barad provides a framework for analyzing how boundaries between the human and the nonhuman emerge and are stabilized. Barad's point here is that apparatuses have no pre-given boundaries, rather they intra-act with other apparatuses and in doing so reiteratively produce what counts as culture and nature, subject and object, active and passive, human and nonhuman, and so forth.

It is crucially important to understand that Barad's reworked notion of the apparatus does not designate static physical "arrangement in the world"; and even less is the term limited on experimental settings in laboratories.

"Apparatuses are not inscription devices, scientific instruments set in place before the action happens, or machines that mediate the dialectic of resistance and accommodation. They are neither neutral probes of the natural world nor structures that deterministically impose some particular outcome. [...] apparatuses are not mere static arrangements *in* the world, but rather *apparatuses are dynamic (re)configurings of the world, specific agential practices/intra-actions/performances through which specific exclusionary boundaries are enacted*." (Barad 2003: 816)

---

42 For a different reading of Foucault's work, one which proposes that the later Foucault tried to find a way not to differentiate between the government of human beings and things, see Lemke (2015). For Lemke, Foucault's notion of a 'government of things' allows for the taking into account of the entanglements of humans and nonhumans, the natural and the artificial, the physical and the moral.

Apparatuses, in Barad's understanding, are therefore neither static arrangements nor are they "external forces that operate on bodies from the outside; rather, apparatuses are material-discursive practices that are inextricable from the bodies that are produced and through which power works its productive effects" (Barad 2007: 230). As both parts of phenomena and as phenomena themselves apparatuses function as "boundary-making practices", determining "what matters and what is excluded from mattering" (ibid: 148). Different apparatuses enact different possibilities. It is also for this reason that measurement, in Barad's understanding, cannot be reduced to particular practices in laboratories. In fact, any "causal intra-action is implicitly a measurement in Barad's sense", as Joseph Rouse (2004: 158, Fn. 8) states. Here, Barad makes clear that the particular (re)configurings that apparatuses take, are nor arbitrary; nor are they constructions of "'our' choosing" or primarily "the result of causally deterministic power structures" (Barad 2003: 829). What is more, apparatuses "may (but need not) include both humans and nonhumans" (Barad 2007: 434, Fn. 65).

To summarize, in Barad's understanding, apparatuses are not so much located somewhere *in* the world but rather specific material-discursive practices that produce differences that matter; that is, apparatuses are sets of "boundary-making practices" (Barad 2007: 146). In Barad's account specifically (re)configured materialities (including material bodies) are intra-actively enacted through particular apparatuses. Apparatuses, however, do not work on the body from 'the outside', as it is the case in so many dystopian stories of technologies and technoscientific practices in which a supposedly natural (meaning, technologically untouched) body is subjugated to and colonized, reprogrammed or even dissolved by the workings of particular (modern) technologies and technoscientific practices. Since apparatuses are prolific material-discursive practices, rather than fixed structures, they do not have intrinsic boundaries.

Barad elaborates this last point by referring to the Stern-Gerlach experiment, performed by Otto Stern and Walther Gerlach in Frankfurt in 1922, in which amongst other things a cigar became a relevant part of the apparatus. The fact that Stern, a notorious smoker, could not stop smoking cigar during the experiment, and the circumstance that the cigars he smoked were of low quality, emitting high amounts of sulfur, was what enabled the trace of the silver atom beam to show up clearer on the detector plate in the first place. Barad argues that since Stern was an assistant professor with a modest salary, he could not afford good cigars, which were simply too expensive in the interwar period. When Stern examined the detector plate at a distance close enough, "the plates could absorb the fumes of Stern's sulfuric breath, turning the faint, nearly invisible. silver traces into jet black silver sulfide traces" (Barad 2007: 165). Barad uses this example to demonstrate that the boundaries of the apparatus go well beyond the actual instrument, incorporating also "class, nationalism, economics, and gender" (ibid: 167) as crucial 'parts' of the apparatus.[43]

Using the example of the ultrasound transducer as an apparatus of observation, Barad further demonstrates what it means to argue that specifically (re)configured bodies (human and nonhuman ones alike) as well as their boundaries are the result of particular material-discursive practices in which not only humans but also nonhumans (including technologies as well as the flows and forces of bodies themselves) take part. The ultrasound or piezoelectric transducer is a device for converting ultrasound waves to electric signals, "information in nonelectrical domains to information in electrical domains", and vice versa (Skoog et al. 2007: 9). Piezoelectricity – from the Greek 'piezein' (πιέζειν) for 'to press' – designates the electric charge that accumulates in certain organic and inorganic material such as, for example, in crystals, skin tissue, and bones, if pressure is applied. As with many other technologies, piezoelectric devices were first applied in a military context as a mean to detect submarines during the First World War. If force is applied, piezoelectric crystals generate a voltage. This property of piezoelectric crystals makes them a crucial component of the ultrasound transducer. Subsequently, Barad uses the ultrasound transducer as a tool for exploring the nature of the relationship between the material and the discursive.

Analogous to Butler's example of gender interpellation at birth, Barad argues that obstetric ultrasonography (as a range of different practices, rather than a single, coherent one) "does not simply map the terrain of the body: it maps geopolitical, economic, and historical factors, as well" (Barad 2007: 194). However, while feminist analyses of reproductive technologies and of medical practices have made evident that not only discursive but also material factors are critical to the process of materialization in the context of obstetric ultrasonography, the question remained open as to how to take account of "the material constraints, the material dimensions of agency, and the material dimensions of regulatory practices" (ibid).

It is precisely the idea that the material and the discursive are always already intertwined, or perhaps better *entangled* with each other, in apparatuses of bodily

---

43 It is important to understand that Barad, here, does not mean that "there are leaks in the system [of proper science, JB] where social values seep in despite scientists' best efforts to maintain a vacuum-tight seal between the separate domains of nature and culture". Rather, the social and the scientific are always "co-constituted", since both are "open-ended, entangled material practices" (Barad 2007: 167-168).

production that leads Barad to the conclusion that "material and discursive constraints operate through one another" (Barad 2007: 212). Rather than "little bits of nature" or "a blank slate, surface, or site passively awaiting signification" (Barad 2003: 821) – that is, "a kind of citationality", as Butler (1993: 15) suggests – matter, in Barad's understanding, is refigured as *materialization*, as,

"substance in its intra-active becoming—not a thing, but a doing, a congealing of agency. Matter is a stabilizing and destabilizing process of iterative intra-activity. [...] That is, matter refers to the materiality/materialization of phenomena, not to an inherent fixed property of abstract independently existing objects of Newtonian physics (the modernist realization of the Democritean dream of atoms and the void)." (Barad 2003: 822)

Reframing matter as materialization, as a process rather than a product,[44] Barad contests the idea of matter as an ahistorical and transcendental substance that is only mechanically following Newtonian laws; an idea that not only runs against the first law of thermodynamics, which states that matter and energy are constantly in transformation from one form into another, but also ignores that this idea of matter "is not a given but a recent historical creation", as Bruno Latour (1999a: 207) states. What is more, understanding matter as "a dynamic play of in/determinancy" (Barad 2012b: 16) evades the problem that,

"Under the rubric of 'matter,' two totally different types of movement had been conflated: first, the way we move knowledge forward in order to access things that are far away or otherwise inaccessible; and, second, the way things move to keep themselves in existence. We can identify matter with one or the other, but not with the two together without absurdity." (Latour 2007b: 139)

Using the example of the ultrasound transducer, Barad elaborates how materialities and meanings simultaneously emerge through particular material-discursive

---

44 It is against this backdrop that in her short essay entitled *What is the Measurement of Nothingness? Infinity, Virtuality, Justice,* Barad puts forward the idea that the vacuum is not 'empty'; while it is certainly also not 'something', but in fact "virtuality", "the indeterminacy of being/nonbeing, a ghostly non/existence" (Barad 2012b: 12). In doing so, Barad goes even so far as to argue that the void is, in fact, "a lively tension, a desiring orientation toward being/becoming", "the infinite plentitude of openness" (ibid: 13: 17). Interestingly, in a recent paper on the expanding of the universe, a similar idea of matter as something that is always contained as potential (or as virtuality) in space is suggested (see Bagchi et al. 2013).

practices. According to Barad, "the transducer does not allow us to peer innocently at the fetus, nor does it simply offer constraints on what we can see; rather, it helps produce and is part of the body it images" (Barad 2007: 202). That being so, the marks on the computer screen "refer to a phenomenon that is constituted in the intra-action of the 'object' (commonly referred to as the 'fetus') and the 'agencies of observation'" (ibid). It is through the intra-action of social *and* material forces and 'components', such as the ultrasound transducer, the computer interface that is hooked up to it, juridical and political norms and discourses, the knowledge and practices of physicians, environmental influences, hormones and genes, and consequently the bodily flows of the fetus 'itself', that specific properties (for example, sex and gender, or dis/ability) and boundaries sediment and become intelligible. These properties and boundaries are therefore neither observation-independent matters of fact nor are they mere social or linguistic constructions. What is more, processes of materialization are not deterministic since apparatuses are "themselves constituted through particular practices that are perpetually open to rearrangements, rearticulations, and other reworkings" (Barad 2007: 203).

Barad's posthumanist understanding of performativity and materialization not only "enables genealogical analyses of how boundaries are produced rather than presuming sets of well-worn binaries in advance" (Barad 2007: 30), but also allows us to take into account the *material* dimensions of agency. Similar to Latour and Haraway, Barad understands agency not aligned with intentionality or subjectivity. Subjects or objects do not hold agency since they do not preexist as such in Barad's philosophy. Consequently, agency is not limited to humans nor is it *extended* to nonhumans because agency is "not something that someone or something has" (Barad 2003: 826-827). Rather, in Barad's account agency denotes an enactment. "*Agency is 'doing' or 'being' in its intra-activity. It is the enactment of iterative changes to particular practices—iterative reconfigurings of topological manifolds of spacetime-matter relations—through the dynamics of intra-activity*" (Barad 2007: 178). In an important sense, agency is therefore "about possibilities for worldly re-configurings" (Barad 2012c: 55). Such a relational account of agency and ontology promises to open up an avenue for a more materialist account of politics[45] or, more precisely, an actualized understanding

---

45 Barad has been accused of not focusing on the politics of gender. Sometimes this accusation comes ironically from male scholars, such as, for example, from Trevor Pinch (2011), as if considering gender and problematizing gender inequalities would be something only feminist scholars have to do or would even be a duty of (and only of) feminist scholars. The circumstance that Barad does not primarily focus on

of biopolitics after Foucault.[46] In any case, such an account demonstrates the need for a more inclusive form of politics, which explicitly contests human exceptionalism. In contrast to Latour's 'parliament of things', Barad, however, argues that, despite the fact that 'we' are always deeply entangled in (and thus part of) a close web of human and nonhuman relations and forces, 'we' are still the ones who are responsible and accountable for what matters as well as for what is excluded from mattering.

Barad's understanding of responsibility is thoroughly influenced by Emmanuel Levinas's notion of ethics. It is particularly Levinas's idea that responsibility denotes a form of 'difference which is non-indifference', what inspires Barad to arrive at an understanding of differentiating as "not about othering or separating but on the contrary about making connections and commitments. The very nature of materiality is an entanglement. Matter itself is always already open to, or rather entangled with, the 'Other'." (Barad 2007: 393) Consequently, as Barad demonstrates, 'we' are always responsible to the Other with whom 'we' are entangled with.

"Responsibility is not ours alone. And yet our responsibility is greater than it would be if it were ours alone. Responsibility entails an ongoing responsiveness to the entanglements of self and other, here and there, now and then. If, as Levinas suggests, 'proximity, difference which is non-indifference, is responsibility,' then entanglements bring us face to face with the fact that what seems far off in space and time may be as close or closer than the pulse of here and now that appears to beat from a center that lies beneath our skin. The past is never finished once and for all and out of sight may be out of touch but not necessarily out of reach." (Barad 2007: 394)

It is precisely for this reason that Barad is very specific in arguing that the fact that there are no agents per se in her approach does not mean that she denies the importance of agency; nor would such an understanding of agency ignore the existence of power imbalances. On the contrary,

---

"women and gender" in her theory does not mean that her approach is not a feminist one. In fact, Barad makes clear that her "concern as a feminist scholar is not women or gender per se, but rather an engagement with feminist understandings of the political. A monocular focus on Politics with a capital P (and only a very particular slice at that) obscures crucial dimensions of politics and power. That is one reason poststructuralism has been immensely important for feminist, queer, critical race, and postcolonial theorists." (Barad 2011: 449)

46 See also Lemke (2015).

"The specificity of intra-actions speaks to the particularities of the power imbalances of the complexity of a field of forces. I know that some people are very nervous about not having agency localized in the human subject, but I think that is the first step— recognizing that there is not this kind of localization or particular characterization of the human subject is the first step in taking account of power imbalances, not an undoing of it." (Barad 2012c: 56)

Instead of restricting power to the domain of the social, Barad emphasizes that power operates through the enactment of "naturalsocial forces". It is also for this reason that Barad locates the "political potential of deconstructive analysis [...] not in simply recognizing the inevitability of exclusions but in insisting on accountability for the particular exclusions that are enacted and in taking up the responsibility to perpetually contest and rework boundaries" (Barad 2007: 205). As reality is neither a fixed essence nor to be understood relativistically in Barad's account it matters what counts as real, what matters, at what costs, and what is excluded from mattering. It is for this reason that rather than understanding ultrasonography as "a surveillance assemblage" (see, for example, Lowry 2004),[47] what Barad offers is an understanding of how particular materialities (along with specific political and ethical consequences) are performatively enacted through ultrasonography as a set of material-discursive practices, that is, as an apparatus of production. And it is precisely this extended notion of the apparatus that promises to be of high value for a new materialist theory of technology and the body.

FIGURATIONS MATTER: THE TECHNO-APPARATUS AS FIGURE AND METHOD

Figures and figurations, that is, the methodological and theoretical use of figures, are, as Donna Haraway (1997: 11) reminds us, "performative images", "condensed maps of contestable worlds" with the power to "trouble identifications and certainties". This is also the reason why Haraway puts the search for new "materialized refigurations" (Haraway 1997: 23) into the center of feminist technoscience studies. Etymologically, the term figuration can be traced back to the

---

47 This, however, does not mean that reproductive technologies, and ultrasonography in particular, are innocent technologies since they very well can embody aspects of surveillance and control. Not to mention that also the question who has access to these technologies (and at what costs) and who does not remains a crucial one.

Latin word *figuratio*, which means 'shaping' or 'forming' and refers to acts of representing something emblematically. And yet, in spite of the fact that figures are images, they do not merely represent or mirror preexisting phenomena. Rather, figures are always also performative and therefore generative, allowing us to make the nonconceptual intelligible. Figurations are materializing technologies that turn bodies into stories, and stories into situated, embodied bodies, "producing both what can count as real and the witness to that reality" (Haraway 1997: 179). That is to say, figures are active modes of thought or, in the words of Rosi Braidotti, "materialistic mappings", fulfilling "the function of providing both analytic and exegetical tools for critical thought and also creative theoretical alternatives" (Braidotti 2011: 4). The generative force and the critical potential of the concept of the apparatus of bodily production therefore lies precisely in "the join between the figurative and the factual" (Haraway 2000: 24). Figures are neither mere metaphors,[48] nor are they simply language games without any material and political dimension. The notion of language games itself is misleading since, following the later Wittgenstein (1953), language games are nothing else than activities rooted in human practices and therefore not cut off from the world and material bodies. Precisely because human practices provide the foundations for certainties to emerge, it is not language that determines reality for Wittgenstein. Rather, language and matter, word and world are always entangled with each another.

In what follows, my aim will lie at outlining an actualized, or perhaps better a re(con)figured, understanding of the apparatus of bodily production. However, as with Haraway's (1992: 298) theorization of the object of knowledge as material-semiotic actor, I do not want to imply an immediate presence of techno-apparatuses of bodily production as actual, discrete objects 'out there'. Techno-apparatuses of bodily production, in my understanding, are neither concrete objects nor necessarily specific spatiotemporally localizable places. Even less can they be regarded as mere material artifacts or machineries in a literal sense that produce particular phenomena – such as, for example, particular bodies and realities. Rather, following the idea that concepts are material (re)configurings (Barad 2007), or *Gestaltungen*, such a re(con)figured concept of the apparatus of

---

48 Haraway insists on the materiality and power of figures by stressing that "words and figures themselves are flesh and do a lot things. But what they can't do is stay still as a conceptual apparatus that makes most philosophers happy, and so they end up saying, 'Mere meta – phor,' and I think, 'Give me a break, guys. This is not mere metaphor, this is actually an enactment of, among other things, corporeal cognitive practice." (Haraway 2016: 277)

bodily production that I propose here refers to both objects of critical inquiry – in the sense of particular sites where biological, technological, social, economic, and political forces intra-act and in doing so mutually materialize a certain phenomenon – and to a speculative tool for engaging with narratives centering on questions of becoming and power in the context of science and technology.

In a certain sense the concept of the techno-apparatus of bodily production resembles much of what C. P. Snow has termed 'deictic object',[49] that is, an object that produces a (material) phenomenon and at the same time is part of that very phenomenon. But it is also akin to what Deleuze and Guattari have termed 'agencement' (or assemblage in its English translation),[50] with which it shares the consequence to dismiss the idea of the human subject as staying at the control-levers, enacting what counts as reality solely through the power of her mind and language. However, while Deleuze states that what exactly defines an agencement "is the AND, as something which has its place between the elements or between the sets. AND, AND, AND—stammering" (Deleuze/Parnet 1977: 34), the question remains how to think this 'AND, AND, AND'; as an addition, as something that connects already given entities with one another, or rather as an entanglement of heterogeneous multiplicities?

"Assemblages operate through desire as abstract machines or arrangements, that are productive and have function; desire is the circulating energy that produces connections. [...] Assemblages emerge from the arranging of heterogeneous elements into a productive (or machinic) entity that can be diagrammed, at least temporarily." (Livesey 2005: 18)

At certain places Deleuze indeed suggests to think assemblages as multiplicities that are "made up of many heterogeneous terms" which establish "liaisons"; or even as "symbiosis" (Deleuze/Parnet 1977: 69). In *A Thousand Plateaus*, Deleuze and Guattari even write that an "assemblage is precisely this increase in the dimensions of a multiplicity that necessarily changes in nature as it expands

---

49 In linguistics, the term deictic (δείκνυμι for "to show", "demonstrate", or "refer to directly") denotes a word that cannot be understood without further context. Examples for deictic words are "I", "the speaker", or "the addressee".

50 Despite the fact that the term agencement is usually translated as assemblage, it has no real counterpart in English. While the term assemblage refers to an assembly, the term agencement denotes the processes of putting together. In an important sense, agency forms a part of agencement, and this is precisely what gets lost in the English translation assemblage. Both as a verb and as a noun, the term agencement does not separate between assemblies or arrangements and agencies.

its *connections*" (Deleuze/Guattari 2004: 8; italics JB). In an important sense, the term assemblage – especially in its original meaning as agencement – denotes an arrangement or a constellation of bodies, objects, technologies, desires, and much more, *and* at the same time also the very process of fitting elements together. For Deleuze and Guattari the elements, or more precisely the forces and flows that constitute an assemblage are not to be understood as stable or self-identical but rather as always in becoming and therefore as never fully detachable from one another. However, while the term assemblage/agencement shares many aspects of the concept of the techno-apparatus of bodily production I am following here the latter is even less concerned about questions of connection and interaction. In fact, the focus shifts from questions of interaction and connection to questions of intra-action and entanglement.

Shifting the focus from given entities that somehow interact with one another to entanglements hence might also rework the notion of reality. Instead of having a fixed essence, reality is reframed as the effect of particular 'cuts' in an amorphous stream of matter-as-practice-in-its-ongoing-becoming. Following the Latin roots of the term apparatus as 'to bring forth something' (apparare; verb), techno-apparatuses of bodily production are then what materialize phenomena locally and temporary out of these streams, lending them a relative stability. As stated before, what constitutes a specific techno-apparatus is never determined in advance. There is no such thing as a list of causes and forces that matter and those that do not, which could be anticipated or excluded in advance. Political, economic, material, biological, technical, and other forces but also social hierarchies and power relations might or might not intra-act within a given techno-apparatuses of bodily production. However, not all of the entities, relations, and forces necessarily *have to be* a constitutive part of a given techno-apparatus at a given time. The only thing that can be known about a techno-apparatus before its analysis is that it consists of a multitude of material-discursive relations and forces and that it is yet able to function toward a certain end, despite the fact that the relations and forces it is constituted might always threaten the techno-apparatus to disaggregate through their attempts to flee. But the question what exactly makes a particular techno-apparatus to what it is in a particular context has to be scrutinized always anew. For this reason, a particular techno-apparatus remains stable precisely as long as it is capable of holding together more forces than those that flee.

It is for this reason that such an understanding of the concept of the techno-apparatus of bodily production not only demands to rethink epistemology, incorporating the objects of knowledge as actively involved in the process of knowledge production, but also of ontology. Instead of having substances with a fixed and static essence, the starting points of such an ontology are relations. Even though matter is reframed as "a doing", and hence as becoming rather than as a "fixed substance" or "a thing" (Barad 2003: 822), such a non- or post-representationalist relational ontology nevertheless remains a materialist one at its heart. As a consequence, being is understood as multiplicity all along. By understanding that "becoming and multiplicity are the same thing", as Deleuze and Guattari emphasize (2004: 275), not only a transcended notion of ontology ceases but also the idea of a unity which could,

"serve as a pivot in the object, or to divide in the subject. There is not even the unity to abort in the object or "return" in the subject. A multiplicity has neither subject nor object, only determinations, magnitudes, and dimensions that cannot increase in number without the multiplicity changing in nature (the laws of combination therefore increase in number as the multiplicity grows." (Deleuze/Guattari 2004: 8)

In this wake, rather than designating eternal stasis, ontology, from a new materialist account of techno-apparatuses of bodily production, is fundamentally reworked as constant becoming-with, as being-as-multiplicity. That being so, the focus shifts from the question of how the world *is*, or even how it ought to be, to question of how the world constantly *becomes-with* and what is excluded from becoming, and consequently from mattering. The fixed modernist ontology as essentially given and therefore as fate is deconstructed and reframed as a specific effect of intra-actions of humans and nonhuman entities and forces – be they organic, technological, discursive, or textual. Moreover, what becomes evident is that different techno-apparatuses enact different phenomena and consequently also different ontologies with different ethical and political consequences, which are not separable from the very techno-apparatuses through which they come into being. If a given techno-apparatus changes also the ontological status of the phenomenon it enacts might change. Thus, what counts as real is neither a mere social construction nor is it identical with an inaccessible preexisting reality 'out there' which would be made intelligible through the power of representation. In any case, the world is not only the product of human practices alone. Because such a relational ontology concerns about questions of mattering, that is, the question under which conditions and through which practices specific (material) phenomena are enacted (and at the same time others excluded from mattering), it is always an explicitly political position. Ontology becomes visible as never solid but constantly in change, and thus as contested ground. Consequently, such a reworked notion of the apparatus of bodily production is loyal to what Donna Haraway calls a more humble, downsized theory. Acknowledging the end of grand narratives (meaning, large scale theories of the world, such as the belief in a linear progress of history),[51] for Haraway, "the production of universal, totalizing theory is a major mistake that misses most of reality, probably always, but certainly now" (Haraway 1991a: 181). The concept of the techno-apparatus of bodily production, employed as a speculative tool for critical inquiries, hence, does not represent a coherent theory or even a Philosophy (with a capital "P") of technology and the body, rather it shifts the focus on question of how particular materialities and with them ethics and politics are enacted through particular technologies and technoscientific practices (from which the bodies materialized cannot be separated just like that).

Understanding techno-apparatuses of bodily production as sites where heterogeneous naturalcultural forces intra-act and at the same time as a speculative tool for investigations into the nature of becoming-with in the context of science and technology not only promises a deeper genealogical understanding of how specific knowledge about bodies *and* specifically re(con)figured bodies – meaning, bodies marked by race, ethnicity, sex/gender, nationality, and even species – in their very materialities are performatively enacted through particular technologies and technoscientific practices. In fact such an account might also allow the taking into account of material bodies as generative and unruly, and therefore as actively involved in the processes of their own reconfiguring, rather than only as mere objects or inert matter on that powerful technologies and technoscientific practices act upon. What is more, by taking up Haraway's question who survives and who dies in technoscience, and at what costs, such a re(con)figured notion of the apparatus of bodily production might also be able to take into account the ever-changing topologies of power – yet precisely without overemphasizing the omnipotence of science and technology, or ignoring the potential of bodies and organisms to be unruly – as I will elaborate in what follows by turning to two "concrete worldly examples" (Haraway/Goodeve 2000: 46): the spirometer, a medical device for measuring vital capacity, and the *Human Provenance Pilot Project*, a project initiated by the UK Border Agency with the aim to combat undocumented migration with new technologies and technoscientific practices such as DNA and isotope testing, targeting the bodies of African asylum seekers.

---

51 See in particular Lyotard (1984).

4 Cutting Technology and the Body Together-Apart

> Entanglements are not unities. They do not erase differences; on the contrary, entanglings entail differentiatings, differentiatings entail entanglings. One move – *cutting together-apart*.
> —Karen Barad/Diffracting Diffraction

In the previous chapters I have argued that social constructionist approaches often fail to theorize the complex, dynamic relationship of entanglement of technologies and material bodies. What is more, they not only fail to grasp material bodies as more than mere passive objects or surfaces on that powerful technologies and technoscientific practices act upon, but in many cases also fail to understand technology beyond a narrow understanding of congealed and condensed social power relations, and hence as a mere placeholder for 'the social'. Drawing on Bruno Latour's philosophy of technology, I have stressed the need for an understanding of technology as a process, or as a movement, in which humans and nonhumans, bodies and things, are inextricably entangled and mutually reconfiguring one another. Instead of understanding technology as the Other of the body, in what follows, I will argue that, in fact, there is no outside to technology. In an important sense, this however does not mean that since technology is always already a part of us there would be no need to explore and problematize specific technological and technoscientific reconfigurings of material bodies. On the contrary, the notion of the techno-apparatus of bodily production, not only emphasizes the need for new critical concepts and figures to emerge, but also explicitly calls for concrete analyses of technologies and bodies in their entanglements with one another with regard to their material, political, and ethical consequences.

As I have argued in the previous chapter, a re(con)figured notion of the apparatus of bodily production as both a figure and a speculative lens promises to be just the right tool for this endeavor, allowing for an understanding of how not only knowledge but also materialities are performatively enacted through specific material-discursive practices – that is, through specific techno-apparatuses of bodily production. As outlined earlier, for Barad, causality is "an entangled affaire", "a matter of cutting things together and apart" (Barad 2007: 394). The complexity at work in the concept becomes visible if it is understood that while 'cutting apart' denotes the processes of differentiation between objects and measurement agencies *within* a given phenomenon, 'cutting-together' refers to the processes of extending the entanglements of a given phenomenon. Cutting technologies and bodies together-apart thus means to trouble the very notion of the dichotomy of technology and the body, yet without constructing a one-ness, because entanglements are not unities, "they do not erase differences" (Barad 2014: 176). Rather, differences and boundaries are reframed as differences and boundaries in-the-making. Following the idea that "diffraction is a mapping of interference" (Haraway 1992: 300), in what follows, I will outline a different story about bodies and technologies, one that goes beyond narratives of loss and domination, without however ignoring the concrete consequences of specific entanglements and processes of materialization.

DIFFERENCE THAT MATTERS:
THE BIOPOLITICS OF THE SPIROMETER

In 1846, John Hutchinson, a British physiologist, built a highly experimental apparatus for measuring vital capacity, that is, the maximum volume of air a person can expel from their lungs after a maximum inhalation:[1] the spirometer. Despite the fact that Hutchinson is usually regarded as the inventor of the spirometer, attempts at measuring lung capacity date back as far as to ancient Greece. With the advent of experimental science and ever more sophisticated technologies, however, apparatuses for measuring the human body and its functions became increasingly accurate and common. It was in particular the work of the French chemist Antoine Lavoisier on the role oxygen plays in combustion that had a great influence on the development of modern apparatuses measuring pulmonary functions at the beginning of the 19th century. In the wake of Lavoisier's work

---

1   Modern spirometry differentiates between two different variables: vital capacity (VC) or forced vital capacity (FCV), and the forced expiratory volume in one second (FEV1). While FEV1 is of more recent origin, vital capacity has a longer history. (Gibson 2005)

Hutchinson not only refined the water-filled spirometer but also redesigned it in order to be able to measure vital capacity in living, breathing people. More importantly however, Hutchinson not only established a direct link between a person's vital capacity and their sex, height, and age, but as a medical assessor for a large insurance company also gathered a vast number of measurements from both healthy patients and from patients with lung diseases, establishing vital capacity as an important variable far beyond the medical area. For the following century and a half, from Victorian Britain to our world today, the spirometer has stood at the center of questions of health, genetics, nationalism, economic interests, race, and social inequality.

In her groundbreaking research on the history of the spirometer, the historian of science Lundy Braun (2014a) elaborates how the biopolitical past of the instrument still haunts the bodies of those marked as deviating from a racialized and gendered norm. In the year 1999, black asbestos workers, who became ill with asbestosis and lung cancer after they had been working in the steel industry for years, confronted their former employer Owens Corning Corporation[2] in Baltimore's circuit court. In contrast to many of their white colleagues, the company refused the payment of disability benefits to the black workers who had fallen ill. As Braun implies, this move on the part of the company was however less an expression of obvious racist bias than it was the result of more stringent standards for black workers compared to their white colleagues, making it harder for them to qualify for compensation in the case of serious illness and disability. As a matter of fact, *Owens Corning* believed "that because blacks score consistently lower on tests used to determine lung capacity, they should have to meet a higher standard to prove asbestos caused lung damage".[3]

---

2   Owens Corning is considered as the world's largest manufacturer of fiberglass. In 2000, the company went bankrupt because of medical liabilities due to the use of asbestos as a fireproofing agent. Six years later, the company emerged from its Chapter 11 bankruptcy. In 2015 and 2016 Owens Corning even received a perfect score of 100 percent on the Corporate Equality index – a national benchmarking survey on corporate policies and practices related to LGBT workplace equality, issued by the *Human Rights Campaign* (the largest LGBT civil rights advocacy group in the US). On its website the company states that they are "pleased to be recognized as an organization where our employees feel included, respected and valued for the great work that, in turn, earns high marks from our customers" (Owens Corning 2015).

3   Los Angeles Times, March 26, 1999, http://articles.latimes.com/print/1999/mar/26/news/mn-21243.

How could it be possible that at the turn to the 21st century people in the United States were treated differently according to race when it comes to compensation payments for disability? How was it possible that black workers had to demonstrate a more serious lung damage than their white colleagues in order to be eligible for compensation payment? Departing from the Baltimore lawsuit, Braun makes explicit that the different standards for the black workers result from the belief that black people would have a lower lung capacity than whites by nature. The denial of compensation to disabled black workers, thus, would had to be regarded as the result of theories about physical and racial difference that date back to at least the early 19th century. It is for this reason that, for Braun, the spirometer stays at the heart of these debates about biological difference and the different worth of human life.

But how could a rather simple instrument such as the spirometer be responsible for this; how could it achieve such a far-reaching social effect? The answer seems obvious to Braun: race and with it social inequality have been inscribed, both intentionally and unconsciously, into the spirometer, producing seemingly objective truths about the nature of the bodies under examination. Engineers and manufacturers program race, or more precisely, 'race correction' – that is, a mathematical adjustment of the data collected, "setting the standard of 'normal' for blacks at a fixed percentage, usually 10-15 percent, below that of a norm based on studies of whites" (Braun 2005: 136) – into the instrument.[4] 'Race correction', means that the measured values for lung capacity and vitality are automatically reduced for patients that are identified as 'non-white', and in particular for black people.[5]

For the last one and a half century the data produced by the spirometer had to be compared with a plethora of numbers and charts in order to be made

---

4   For a more detailed medical discussion on the question which correction factor has to be applied to whom and when see, for example, Ruppel (2006: 3) who states that the 'race correction factor' is applied "to Caucasian-derived predicted values to make them appropriate for other races (0.85 to 0.88 for African-Americans)".

5   In an important sense, race, in the majority of the cases, is not about self-identification here but extrapolated from phenotype, or even skin color. The reason for this is, as Braun (2014b) argues, that "it can be hard to ask someone their race for a lung function test. Patients might wonder why race is relevant for this particular test. So, in general, my research suggests that operators/clinicians simply guess a patient's race based on the usual simplistic physical characteristics historically associated with 'race,' like skin color—a poor marker for race globally. This guess may have little to do with how someone self-identifies or the richness of their ancestry".

meaningful. However, this changed with the shift from water-seal volume displacement spirometers to electronic flow measurement machines in the early 1970s. With the availability of cheaper and faster microprocessors, analog instruments were quickly replaced by digital ones which were not only smaller and thus more transportable but made it also less necessary to dig into charts and statistics in order to make the date gathered meaningful. While the machines of the 1980s and 1990s often had a switch for 'race', amongst other categories such as 'sex' and 'height' (Figure 1), contemporary spirometers look much like small, portable computers with touchscreen, and cleared of any obvious links to categories such as race.

Even though not only the shape and the appearance of the instrument has changed since its invention more than one and a half centuries ago, but also the algorithms implemented into the software of the instrument render it often needless to compare the data produced to statistical charts or to use complex mathematical formula in order to make the data gathered meaningful, what remained historically contingent is the idea that people of color, and in particular black people, would have a lower lung capacity and thus a lower vitality than their white counterparts. It is precisely for this reason that, according to Braun, race has been not only embedded into the hardware of the spirometer but also, with the shift from the analogous to the digital, into its software, rendering the spirometer a powerful measuring instrument, a technology into which race had been 'breathed into'. In best social construction of technology tradition, Braun shows that the acceptance of innovation "is not simply a question of its technical superiority or its scientific uses" (Braun 2005: 137); rather, new technologies have to be considered as successful if they are accepted by relevant social groups.

Braun provides a highly insightful historical analysis, taking into account the social and economic interests as well as ideological beliefs that are embedded into the spirometer as a technical instrument. The spirometer, in this light, not only represents an important medical device but also a tool that has been used as an instrument of oppression in the history of the US. But what would it mean to understand the spirometer differently, namely as an apparatus of bodily production rather than as a mere technical object? What would it mean to understand the spirometer less as an object and more as a site where the biological, material, political, economic, and technological fold into one another, enacting specifically reconfigured bodies, which are not separable from the very apparatus through which they come to matter? Would such an account allow for a deeper understanding of how not only particular meanings but also specific bodily materialities along with far-reaching ethical and political consequences are performatively enacted? What is more, would such an account provide us with an

Figure 1: Two electronic spirometers from the late 1980s/early 1990s. A Riko Spiromate AS-300 (above) and a Welch Allyn Pneumocheck 61000 (below). Both devices have a control switch for 'race correction'.

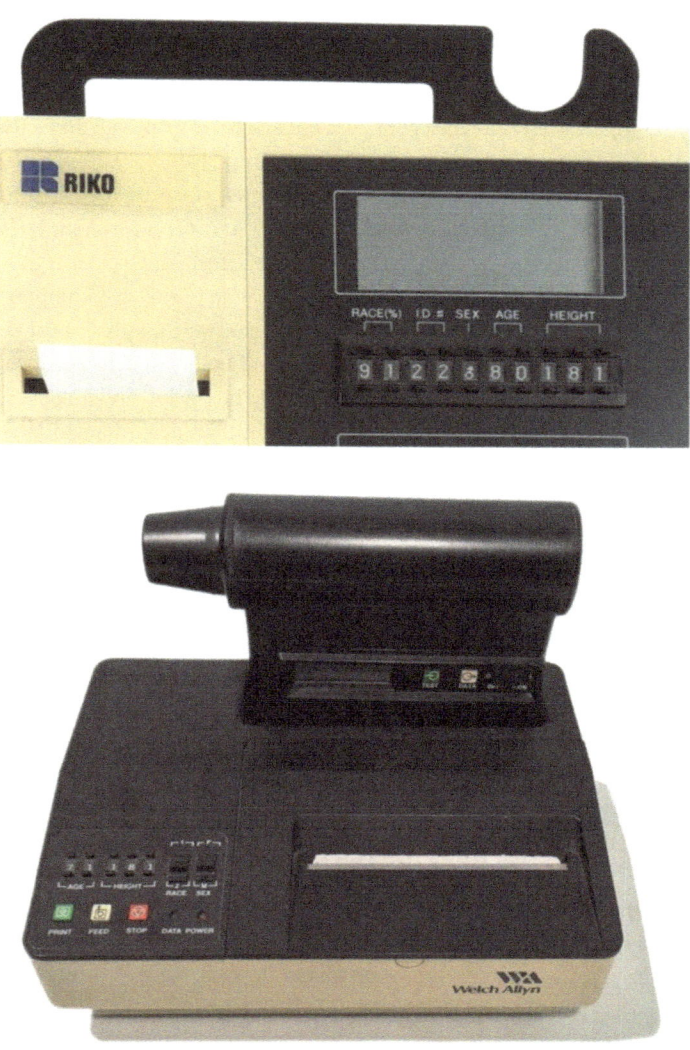

Source: © Ines Handler

understanding of how the concept of vital capacity itself acquires materiality as a complex embodied entity that cannot be separated from the very apparatus of bodily production through which it is enacted?

Braun's study on the history of the spirometer offers most of what is needed to understand the spirometer as a techno-apparatus of bodily production. Taking into account the spirometer as a techno-apparatus of bodily production means to depart from the idea that, rather than understanding it as a technical object with clear and defined boundaries, there is no way of determining in advance where a specific apparatus ends, because apparatuses are "open-ended material-discursive practices" (Barad). Similar to Bruno Latour's notion of the collective, or Gilles Deleuze and Félix Guattari's assemblages as entangled multiplicities, techno-apparatuses of bodily production lack homogeneous closing elements that would allow for a clear-cut differentiation between 'inside' and 'outside'. This does not, however, mean that there are no borders at all or that determining the borders of a concrete techno-apparatus would be an arbitrary task; and that ultimately everything matters to the same degree. Since apparatuses are open-ended practices with no a priori determinate borders "any particular apparatus is always in the process of intra-acting with other apparatuses" (Barad 2007: 203). Hence, it is crucial to examine what matters, how it matters, and for whom it matters.

To understand the role of the spirometer as a technical object staying at the heart of debates about nature and culture, racism and social inequality, biopolitics, nationalism, and the political economy, it is necessary not only to take into account prevalent social, technological, economic, political, and scientific developments that occurred in the mid-19th century but also the new source of power that emerged at the same time, which Michel Foucault has termed 'biopower'. As Foucault explains, the formation of biopower goes hand in hand with the rise of the modern nation state. As a new technology of power, biopower targets life itself, rendering it an object of power Biopower, thus, can be understood as the force that "brought life and its mechanisms into the realm of explicit calculation and made knowledge-power an agent of transformation of human life" (Foucault 1978: 143). As a consequence, biopower gave rise to a new form of politics, namely biopolitics, aiming at the control over the population as a whole – that is, over the entirety of individual physical bodies and over the collective political body. This new biopolitics of the population "gave rise to comprehensive measures, statistical assessments, and interventions aimed at the entire social body or at groups taken as a whole", as Foucault (1978: 146) states. Two centuries after Thomas Hobbes had come up with the image of the state as *Leviathan* – originally, an enormous and powerful sea monster from the Old Testament, in

Hobbes however a metaphor for the absolute sovereign – functioning as counterpart to the state of nature, the idea of the nation as an entity, as a quasi-organism, once again became foregrounded. Individuals merged into populations governed through the means of science and technology. Statistics became an instrument for reshaping the nation's body for the needs of the political economy. It is for this reason that even though biopower rests clearly on an economical base, it is important to understand that for Foucault it cannot be reduced to economics.

In the first half of the 19th century, England, like many other European countries, was shaped by the change from agricultural to industrial economy. In the wake of the development of the steam engine, new technologies in transportation such as railways as well as faster and more efficient ships along with new technologies of production emerged. The soviet physicist Boris Hessen (2009 [1931]) implies that this development that had happened by the turn to the 19th century was to a lesser extent the direct effect of the application of new scientific knowledges but rather an answer to prevalent social and economic challenges. Early industrial capitalism needed new technologies to maximize the efficiency of machines and, at the same time, to reduce the costs of production and transportation. The steam engine can be seen as a technology that initially was employed in mines to pump out water from mineshafts. However, since early steam engines were not powerful and reliable enough,[6] their use was rather limited. According to Hessen, the economic need to increase production demanded an increase in the power and efficiency of the steam engine, before its use could be extended to other economic areas. To accomplish this it was important to get an idea of the concrete forces and thermodynamic processes working 'within' the machine, which were not fully understood at that time. It was only then that James Watt came into play, taking on the mission to elaborate the processes and laws that work within the steam engine in order to be able to develop better, meaning, more economically efficient models. In this light, following Hessen's idea on the emergence of technological innovation, the fact that James Watt laid the foundations of thermodynamics has to be regarded as the consequence of

---

6 Andrew Feenberg (2003) points out that steam engines, or more precisely, steam boilers often blew up in the early 19th century. This was a problem that particularly concerned steamboats and locomotives. After a number of devastating accidents with many casualties, steamboat boilers became the first technology regulated by the US government. In this case, the law, but also increasing economic losses, made it necessary to re-engineer a given technology: the steam engine.

specific industrial and economic needs, rather than the work of a scientific genius.[7]

New Technologies soon greatly increased productivity. However, this socio-economic development came at a high price. Despite the fact that with the advent of bacteriology, new kinds of medical treatments and possibilities to immunize against potentially deadly diseases gained currency, nevertheless many people regardless of social class suffered from infectious diseases such as syphilis, cholera, and tuberculosis. The advance of industrialization had severe social, economic, and ecological consequences not only for the nascent working class, who suffered from long working hours and low wages, as well as from dangerous and unhealthy working places, but for everyone else. As a matter of fact, no one was left unaffected from the far-reaching social, economic and ecological changes that took place in many European cities in the mid-19th century. The inhabitants of large cities such as London and Manchester had to struggle with catastrophically unhygienic working and living conditions. Reports from the second half of the 19th century, for example, describe the Thames as a viscous, brown broth of disgusting smell. From this period date a number of famous drawings and wood engravings such as William Heath's "Monster soup commonly called Thames water" from 1828 (in which an elderly woman holds a microscope magnifying a drop of Thames water, only to reveal that it is filled with all kinds of grotesque aquatic creatures and germs, dropping her full cup of tea) (Figure 2), or "The lamentation of Old Father Thames" from the 1850s (showing a bearded old man surrounded by dead human and animal body parts drifting in the water), depicting the polluted river Thames as a treacly brown foul-smelling fluid filled with sewage, large amounts of rubbish, factory waste, excrements, and offal from slaughter houses. Not only the water and the soil were heavily polluted at that time but also the air. Smoke-belching factories transformed the air into smog and lead to the rise of death rates due to air pollution and respiratory diseases.[8] At the same time, scientific theories on heredity along with cul-

---

7   Even though Hessen provides us with a powerful critique of the idea of the sole male inventor by taking into account social, economic, cultural and many other factors that were involved in the development of a particular theory or technology, he is also running the risk of exchanging technological determinism for a social determinist understanding of science and technology.

8   For a detailed historical study of the relationship between industrialization, filth, pollution, and disease see Lee Jackson's (2015) excellent history of sanitation in Victorian London.

Figure 2: "Monster soup commonly called Thames water, being a correct representation of that precious stuff doled out to us!!!" satirical print made by the British artist William Heath (1828).

Source: © Trustees of the British Museum. All rights reserved.

tural narratives on evolution and progress were gaining ground. Eugenics emerged as an academic discipline, propagating the imagined 'degeneration' of (parts of) the population. It was against this background that the idea of the nation as a social body that would be threatened by an alleged moral, physiological, and cognitive 'degeneration' became more and more popular amongst statesmen, physicists, and philosophers alike. However, for many proponents of this idea, the condition of the national body could be 'improved'. In the wake of the emerging idea of the nation as a quasi-organic body, ideas of controlling, fostering, and optimizing the population flourished in many European countries of the 19th century.

Emerging from attempts at fixating measurement standards and with them difference and variability, statistics increasingly became a powerful tool for realizing particular political ideas and goals.[9] The spirometer was one of the instruments of choice for the collection of the data that was needed for these endeavors. While the spirometer produced purportedly truths about bodies and their capacities,[10] statistics made the data collected meaningful. Ian Hacking emphasizes that something important happened in the first half of the 19th century that

---

9   It is precisely for this reason why the history of modern statistics, the emergence of the insurance industry and insurance mathematics, the advent of epidemiology and the hygiene movement, and the rise of demography as the study of populations, as technologies of governing both the individual and the collective body, cannot be separated from the birth of biopolitics.

10 In an important sense, it was, and yet it was not, the spirometer that produced true statements about 'nature' in those debates. As Donna Haraway illustrates drawing on Steven Shapin and Simon Schaffer's highly influential work *Leviathan and the Air-Pump: Hobbes, Boyle, and the Experimental Life* (1985), the prevalent belief until well into the 19th century was that scientists do not produce facts. Rather, it is Nature (with a capital N) that 'acts', and the instruments were seen as a mean to differentiate between true and false claims about the world. Consequently, the role of scientists was seen to witness and act as Nature's spokesmen, lending the measured data authority. As the facts spoke for themselves, the scientist was nothing more, but also nothing less, than "the legitimate and authorized ventriloquist for the object world, adding nothing from his mere opinions, form his biasing embodiment. And so he is endowed with the remarkable power to establish the facts. He bears witness: he is objective; he guarantees the clarity and purity of objects. His subjectivity is his objectivity. His narratives have a magical power—they lose all trace of their history as stories, as products of partisan projects, as contestable representations, or as constructed documents in their potent capacity to define the facts." (Haraway 1997: 24)

reconfigured biopolitics crucially. A "sharp change" occurred "in the two decades 1820-1840", constituting "the era of enthusiasm for statistical data-collection" (Hacking 1982: 281). This "avalanche of printed numbers", as Hacking (ibid: 290) terms it, can be characterized by the application of statistics for assessing and consequently for attempting to control the lives of the population in almost every aspect. Statistics as scientific method became a tool for campaigns and projects aiming at the 'improvement' of the social body. Since "counting" is "hungry for categories" (ibid 1982: 280), soon ways of not only counting but also of categorizing individual bodies along sex, dis/ability, occupation, and social class emerged; and it did not take long until also race came in as an essential category.

It was against this backdrop that Hutchinson employed the spirometer as "a precise and easy method of detecting disease", as he puts it already in the title of his treaties on the spirometer in 1846. The main function of the spirometer, however, was not only to diagnose pulmonary diseases in order to be able to treat the people who had fallen ill, but also to monitor the physical fitness or vitality of the population, or more precisely, the fitness of working class males since females had been rendered invisible by testing only male bodies. While male bodies functioned as the unmarked norm, female bodies embodied abnormality. It was in this light that Hutchinson's spirometer allowed demonstrating that pulmonary diseases such as tuberculosis lead to a significant decline in vital capacity. Moreover, in the wake of increasingly disastrous effects of industrialization on the bodies of the working class, Hutchinson's spirometer helped, at least indirectly, to lend strength and authority to the belief that it is "foreign matter", that is, germs, bacteria, and dirt in the air and water that, despite not being visible, affects our health (cf. Bishop 1977). Thus, in an important sense, Hutchinson's spirometer represented the instrument of choice for testing and selecting 'productive' bodies from the 'unproductive' ones. It is precisely in this sense that a Mr. Fisher replied to Hutchinson after his talk in front of the Statistical Society of London in the year 1844 that "he had in his lifetime examined 140.000 men for public duty, and had he possessed such an instrument as Mr. Hutchinson's, it would have saved him many times from selecting men who afterwards turned out not eligible for duty" (Bishop 1977: 385). Even though Fisher refers here to police officers and soldiers, the aim to discipline bodies in order to make them more effective for the economy and consequently for the nation, was true for the entire working class body, since labor power was believed to equal national prosperity and consequently national power. Breathing capacity soon became an important indicator for the vague notion of bodily efficiency.

What would it mean now to understand the spirometer, rather than being a mere technical instrument, as a techno-apparatus of bodily production? That is, as a 'site' where material-discursive re(con)figurings of bodies occur, along with far-reaching social, economic, and political consequences for the bodies reconfigured. Even though I am using the same word here as Braun does, namely 'spirometer', I shall not talk about the mere instrument but rather about an assemblage of technical, medical, scientific, economic, and political entities, relations, and practices. As such, the spirometer not only sheds light on how the idea of (biological) vitality and (economic) productivity were entangled by the turn to the 20th century, and that they still remain entangled until today, as the Baltimore lawsuit demonstrates, but also emphasizes how deep matters of public health, hygiene, and sanitation were intertwined with political ideologies and beliefs (and still are). Only in this regard can it be understood how the spirometer amongst other technologies, practices, and discourses helped in materializing the idea of a national body. This function of the spirometer at the intersection of the political economy, scientific theories on evolution, an imagined 'degeneration' of individual bodies and the body of the nation, and heredity, as well as discourses on the worth of human life becomes even more apparent if it is also taken into account that John Hutchinson was not only a physiologist but also a consultant for the *Britannia Life Assurance Company* located in London. As a matter of fact, Hutchinson believed that vital capacity could and indeed "should be used in actuarial predictions for persons selling life insurance" (Petty 2002: 219).

"Before I conclude, I may venture to draw the attention of those connected with insurance offices to the matter of this paper. Thus, the state of a man examined, and appearing like the three first cases in Table X., would admit of little doubt but that such was an assurable life, while the other cases would be suspicious. From such a table of facts, any man can form his own judgment of a case, without being dependent on the opinion of another. Therefore, to insurance offices, as well as to the medical profession, the Spirometer, I think, would be found useful." (Hutchinson 1846: 247)

In his short treatise *The Spirometer, the Stethoscope, & Scale-Balancing; Their Use in Discriminating Diseases of the Chest, and Their Value in Life Offices With Remark on the Selection of Lives for Life Assurance Companies* from 1852, Hutchinson further elaborates this idea, providing dozens of charts for underlining the usefulness of the spirometer for the emerging assurance industry.

The fact that the insurance of persons in Europe and the US thrived by the end of the 19th century was no coincidence. Mechanization and industrialization as well as highly unsafe conditions of work increasingly threatened working class bodies. Low wages, however, only allowed for a modest life at best and were sometimes even not enough to cover the costs of a proper burial. In the need of mutual support, many workers in England and other European countries formed groups for purposes of self-insurance. Soon, however, these small groups became a thorn in the flesh of capitalists and politicians who suspected that "the friendly societies served as a cloak for illicit combinations, i.e. trade unions" (Hacking 1982: 283), which were banned and considered illegal at the time. Taking this socio-historical background into account makes explicit why, by the late 19th century, the insurance industry was "among the biggest, fastest-growing, and most aggressive corporate entities in existence" (Wolff 2006: 87).

Where quality comes in, hierarchy is never too far behind. It is in this sense that in the wake of Hutchinson's work, *difference* increasingly became represented as *deficiency*. While Hutchinson did not attach value to race, it was another statistician and employee of a huge US based insurance company, Frederick Hoffman, who attributed an explanatory function to race in questions with regard to vitality. Hoffman was a German immigrant and one of the leading statisticians at the turn to the 20th century who combined statistics and mathematics with social-Darwinist beliefs as well as racist and white supremacist ideology (cf. Wolff 2006).

After slavery had been formally abolished in the US, the decades after the American Civil War were dominated by debates focusing on both questions of the emancipation of black people in the US, and on attempts of the pro-slavery lobby to scientifically demonstrate that black people were not made for an autonomous life in freedom. Physicians and scientists who often also happened to be plantation owners argued that science would prove that the bodies of people of color, and in particular of black people, were 'biologically not suitable' for freedom. Following Achille Mbembe it could be said that what becomes visible here is how and why slavery and forced labor can be regarded as "one of the first instances of biopolitical experimentation. In many respects, the very structure of the plantation system and its aftermath manifests the emblematic paradoxical figure of the state of exception" (Mbembe 2003: 21). The loss of the rights over one's own body in slavery leads Mbembe to the conclusion that "slave life" has to be regarded as "a form of death-in-life" (ibid). It is this power over the life, or, perhaps more precisely, the death, of enslaved bodies that Mbembe terms as necropower. Hence, necropolitics designates nothing less than the subjugation of life to the "power of death" (ibid: 39).

The attempts to scientifically and technologically prove the lesser worth of black bodies and lives were, however, of course not purely ideologically but also socio-economically motivated. For plantation owners, slavery was an important source of wealth. Frederick Hoffman built on these beliefs when he applied the spirometer, in combination with cutting-edge statistics, to scientifically reinforce the belief in the purportedly inferiority of black bodies. For Hoffman, the spirometer allowed the production of 'objective facts', that is, 'truths', about the bodies analyzed, 'proving' that black people had lower lung capacities than whites. Linking lung capacity to vitality, Hoffman, eventually, came to the conclusion that black people are not only physiologically distinct from white people by 'natural law' but also differ in quality (cf. Braun 2005).

Hoffman, however, was not only a fierce advocate of white supremacy but as mentioned before also an employee of one of the largest insurance companies of the world at that time. Hence, Hoffman's work cannot be separated from the fact that at the beginning of the 20th century "the life insurance industry was expanding in both economic and social power" (Wolff 2006: 84). The circumstance that black people in the US entered the labor marked as free persons, however, brought the insurance industry difficulties. Due to poor living conditions, black people had not only shorter life expectancies than their white counterparts but also limited access to institutionalized education and health care. As unskilled workers, many black people had to work for low wages under highly dangerous and life threatening conditions. For the insurance companies, selling insurances to people of color, therefore, was considered as not profitable. It has to be understood in this sense that insurance companies asked for scientific evidence that would 'objectively' illustrate how human bodies differ in quality according to race; meaning, how black bodies would be inferior by nature. Hoffman and others provided these data with the help of the spirometer, allegedly proving in the wake of eugenics, social-Darwinism, and racial hygiene, that the organs and the bodies of black people were less efficient compared to those of their white counterparts. Considering black workers as clients, hence, would had meant to downgrade the standards for everyone, with the effect that the insurance industry would had to expect financial losses. Megan Wolff puts it in a nutshell, saying that the imagined "biology of African Americans stood in the way of progress; as a race they were antithetical to the goals and rhetoric of commercial insurance" (Wolff 2006: 86). From this followed the purportedly technologically and scientifically informed conclusion that people of color (and in particular, black people) have to be considered as 'uninsurable'; or at least the insurance fees for people of color had be raised in order to make it profitable for insurance companies to sell their products to people of color. It is this idea, embodied in the spirometer, that still haunts black bodies and founds its continuation in the Baltimore lawsuit at the turn to the 21st century.

It is no exaggeration to argue that the spirometer lies at the intersection of science, technology, economics, and biopolitics. What constitutes the spirometer as an apparatus of bodily production is nothing less than the entanglement of the political economy and burgeoning capitalism, the emergence of anthropometry as a discipline, genetics and epigenetics (in form of the effects of environmental influences on the expression or suppression of genes, affecting protein synthesis), the interests of the life insurance industry, the belief in objective knowledge that is produced by scientific instruments, the flows and forces of bodies themselves, as well as the androcentric and ethnocentric idea and practice of setting bodily functions of white males as the unmarked norm. What is more, the spirometer as a techno-apparatus of bodily production makes explicit that the formation of the modern nation state in the 19th century cannot be separated from the emergence of statistics, the insurance industry, and modern 'scientific' racism.[11]

However, the enthusiasm for quantification and the knowledge that "derived with the specialized instruments of the laboratory and organized according to systems of classification" (Braun 2005: 168) in the mid-19th century, as well as the fact that the spirometer cannot be separated from the historical entanglements outlined above, while true is only half of the picture. The spirometer certainly is a technology of "surveillance and discipline", as Braun (2005: 138) argues; a tool "for managing bodies" (ibid: 146). But in an important sense the spirometer is much more than that; it is also a techno-apparatus of bodily production, a generative node, through which specifically reconfigured bodies are hailed into being, not only as socially and discursively constructed bodies, as bodies whose surfaces have been 'overwritten', but also as *materially* reconfigured bodies. The spirometer does not map the terrain of the body, rather it maps biological, social, technological, economic, environmental, and many other human and nonhuman forces and factors. Hence, the spirometer does not only function as a mapping device but also as a site for the making and remaking of bodily boundaries, properties, qualities, and hence biopolitics. The spirometer, as a techno-

---

11 For Foucault, 'scientific' racism is nothing less than "a way of introducing a break into the domain of life that is under power's control: the break between what must live and what must die. [...] It is in short, a way of establishing a biological-type caesura within a population that appears to be a biological domain" (Foucault 2003: 254-255). It is precisely in this light that Thomas Lemke emphasizes that 'scientific' racism is not an irrationality or a terrible mistake in the formation of nation states. In fact, quite the opposite is true as "racism guided the rationality of state actions; it finds form in its political instruments and concrete policies as 'State racism'" (Lemke 2011: 42).

apparatus of bodily production, not only produces the entity it is measuring, that is, 'vital capacity', but also performatively enacts the corresponding bodies, marking black bodies, due to their purportedly lower vitality, as inferior by nature compared to white ones. Understanding the spirometer as an apparatus in this sense, thus, not only sheds light on the material-discursive practices through which race became (and still becomes) embedded into the hard- and software of technical instruments, but also on the practices through which marked bodies emerge as deviation from a (white, male, and able) norm.

Furthermore, what becomes evident from such a point of view is that vital capacity is not something that can be found somewhere 'out there'; but neither does this mean that it is a mere *social* construction. Rather, vital capacity manifests as the material-discursive effect of intra-acting forces and practices (bodily, economic, political, technoscientific, and technological ones) as well as beliefs and ideologies such as nationalism, racism, classism, notions about what a natural, that is to say, a 'normal' and healthy body is and how it ought to function, and much more. In considering the spirometer as a techno-apparatus of bodily production in the sense of an entanglement or a site, rather than a proper object, not only a more materialist understanding of performativity is provided but also demonstrated that measurements have both epistemological as well as ontological, and consequently also social and political consequences. Who is eligible for compensation payment and who not? Whose lives matter and whose not?

While the spirometer clearly belongs to the paradigm of discipline and biopower, the phenomenon I shall turn to next already has to be located into the paradigm of control and technobiopower. Technobiopower, as Donna Haraway has put is so aptly, is to a lesser extent concerned with "dramas of health, degeneration, and the organic efficiencies and pathologies of production and reproduction", but rather with the implosion of "the technical, organic, political, economic, oneiric, and textual" (Haraway 1997: 12), and the multitude of new entanglements that unfold in between.

MATERIALIZING AUTHENTIC BODIES: THE *HUMAN PROVENANCE PILOT PROJECT*[12]

In his *Postscript on the Societies of Control* (1992), Gilles Deleuze stresses that in the late 20th century what Foucault had described as disciplinary societies, with their technologies of punishment and enclosure in the shape of factories, prisons, and asylums, paved the way toward societies of control. Societies of control do not any longer function primarily through detention and punishment, but rely on relentless communication and control. Most importantly, while discipline societies can be characterized, following Foucault's famous turn to Jeremy Bentham's image of the Panopticon, in terms of a fixed architecture,[13] where it is possible to see everything from a central location, there is much to suggest that in control societies there is no such thing as an all-seeing eye in the center that would allow for observing everything and everyone. Rather, surveillance, and consequently, control take place from multiple sites at the same time, and without much restriction as to time and space. For this reason it appears that, despite its popularity, the figure of the Panopticon might has outlived its critical potential to some extent. If control today is hard to trace and localize, more of a

---

12 Parts of this chapter have been previously published as "Technologies of Failure, Bodies of Resistance: Science, Technology and the Mechanics of Materializing Marked Bodies" in *Mattering. Feminism, Science, and Materialism*, ed. by Victoria Pitts-Taylor, New York University Press, 2016, and as "Technoecologies of Borders: Thinking with Borders as Multispecies Matters of Care" in *Australian Feminist Studies*, Vol. 32, No. 94, Winter 2017.

13 Even though the Panopticon is usually regarded as the intellectual property of Jeremy Bentham, it was his brother Samuel who, in fact, mentioned the Panopticon first and also enclosed some drawings of it in one of his letters from Russia in the year 1786. Serving as a military engineer for Prince Potemkin in the war against the Turks, Samuel Bentham witnessed the construction of a Panopticon as a workhouse made of wood in the shape of a star so that every corner could be watched from a central position (see Bentham 1995: 33-34). It was only years later that, influenced by his brothers descriptions, Jeremy Bentham wrote a letter including a diagram of the building his brother saw in Russia to the British Government with the hope of getting sufficient funds for building a factory in the shape of a Panopticon. Jeremy Bentham did not manage to get his Panopticon built. Similar buildings however were constructed from the early 19th century to the first half of the 20th century all over the world. Perhaps the most famous one was the Presidio Modelo penitentiary on Isla de la Juventud in Cuba, built in the late 1920s.

gaseous phenomenon rather than a fixed structure, decentralized rather than having a solid center,[14] the figure of the Panopticon might not be the best choice for analyzing these changed social and technological relations adequately.[15]

While discipline societies were shaped by solid machines involving energy, such as, for example, the steam engine, societies of control are characterized by much more fluid technologies such as computers, cybernetics, and biometric technologies. Science and technology take on a crucial function in societies of control. Foucault elaborates in great detail how disciplinary enclosure operates as an instrument for adapting individuals and their bodies to the needs of the political economy, incorporating them into the capitalist apparatus of production. Enclosure is about aligning and concentrating bodies, it is about composing "a productive force within the dimension of space-time whose effect will be greater than the sum of its component forces" (Deleuze 1992: 3). Discipline does not disappear with societies of control since the political economy still relies on living, working bodies in order to function. However, in an important sense it is not so much the utilization of the collective body for the needs of the political economy but the modulation and surveillance of the individual body that gains center stage. As a new mode of power, modulation on a molecular level (which replaces the disciplining mold that produces docile bodies) directly targets the flows of the body. Where discipline societies target the masses as "a collection of separated individualities" which are "merged together" (Foucault 1977: 201), control societies reconfigure, or remodulate, as Deleuze says, the individual. What is more, individuals themselves become "'*dividuals*,' and masses, samples, data, markets, or 'banks'" (Deleuze 1992: 5). This becomes perhaps nowhere

---

14 This, however, does not mean that there are no such things as 'centers of calculation' (see Latour 1987) where information flows intersect, concentrate, and, in doing so, gain weight.

15 An alternative to the concept of the Panopticon can be found in Bruno Latour and Emilie Hermant's (1998) notion of oligoptics. In contrast to Bentham's and Foucault's megalomaniac Panopticon, oligoptica have "extremely narrow views," but what they see, they see well (Latour 2005a: 181). However, to achieve the anticipated effect a number of human and nonhuman actors must be enrolled, aligned, and kept together. That being so, as rhizomatic networks consisting of humans and nonhumans, oligoptica are far from being powerful technologies against which any resistance would be futile. Quite the contrary, oligoptica are fragile and vulnerable. Even "the tiniest bug can blind oligoptica" (ibid). Turning away from megalomania and paranoia, such a perspective leaves considerably more room to think the possibilities for nonconformity and opposing practices.

more obvious than in the context of genomics and biometric technologies. Surveillance here is not only about gathering information about the individual and its body with the help of certain technologies, but rather about techno-scientifically "decoding and recoding it, sorting it, altering it, circulating it, replaying it" (Bogard 2006: 108). Biometric technologies are characterized by their ability to dissect the individual's body into the totality of its genes, the elements that are incorporated in it, its biochemistry as well as its reactions to environmental influences. In an important sense, however, as I will demonstrate in what follows, biometric technologies not only dissolve but also enact particularly reconfigured bodily materialities.

The term 'biometrics' derives from the Greek words bios (βίος) for life and métron (μέτρον) for measure. Biometric technologies function by collecting information about the body, or more precisely about parts of the body, and translating this it into mathematical variables. Biometric technology, however, is not a particularly new technology. In a certain (analogous) form, technologies of identifying individuals by measuring their bodies or parts of them (for example by taking fingerprints) have already existed for some centuries. In general, biometric technologies can be used in two different ways: for identification (in this case someone's identity is revealed by comparing a measured biometric against a database) and for authentication (here the goal is to determine whether a person is authorized to do or to possess something by comparing a measured biometric against information gathered from the same person at some point in the past). Hence, it is in this sense that biometrics is always and necessarily caught in the paradigm of surveillance, authentication, and representation, and therefore in a mode of thinking that only knows the original and the (illegitimate) copy. Biometric technologies comprise of a number of instruments and machines such as scanning devices that capture the matter and the information flows of the body, software that translates the collected information into a digital form by using algorithms to render it meaningful, and databases in which the information is stored and from where it can be accessed.

Operating from the premise that bodies do not lie, biometric technologies are increasingly deployed for monitoring and controlling migration flows and the movement of individuals. In the early 21st century borders are progressively becoming technologized borders. Satellites surveying suspicious movements from their orbit, illuminated high-security borderlines between states that can be seen from space, drones spotting small boats on the sea, and automated turrets guarding the border,[16] to name but a few, seem to suggest that borders have become technological borders. Even though it could be argued that borders have always been technological, from the first marches and wooden walls to protect Neolithic settlements to the high-tech, barbed-wired fences today, not only the quality and intensity of the technologization of borders but also their form and function has historically changed. Today, perhaps more than ever, the clear lines that traverse our globes and maps seem to give us the impression that borders of nation states are quasi-natural and timeless.[17]

---

16 In 2007, the Israel Defense Forces (IDF) started stationing automated gun systems in pillboxes along the Gaza border. Recently, the stationary guns were accompanied by the unmanned observation security vehicle *Guardium*, the six-wheeled military robot RoBattle, and the unmanned ground combat vehicle *Avant Guard* equipped with a ground penetrating radar, thermal surveillance cameras, and modules for automatic weapons. The IDF's plan for the near future is to establish so-called "automated kill zones" that are guarded by a network of "remote-controlled machine guns, ground sensors, and drones along the 60-kilometer border" (see Shachtman 2007). While these systems are still at least partially remote controlled, the "idea, ultimately, is to have a 'closed-loop' system — no human intervention required" (ibid). Such a project raises a number of serious ethical and political questions such as for example, the question who is accountable and responsible for the killing of people if this represents an automated process in which no humans are actively involved. In contrast to the use of drones for example, in this case there is no individual, not even one that is sitting in an air-conditioned office complex at the other end of the world, who pulls the trigger but rather specific algorithms 'decide' who ought to be killed under which circumstances.

17 As a matter of fact however, state borders are a rather recent development. Until the early 17th century and the invention of instruments and new cartographical techniques necessary for producing more accurate maps most states did not recognize their neighbor's sovereignty over their territory. It was only in the wake of the Peace of Westphalia that borders in their modern form and meaning began to emerge in Europe, along with a naturalized understanding of territory that would function as reference point for images and imaginations about origin, kin, belonging, and community. Historically, the phenomenon of the border went by different names and meanings. The march, the boundary, the frontier, the fence, and the wall are but a few names for distinct phenomena that nevertheless share something common — namely, the introduction of "a division or bifurcation of some sort into the world" (Nail 2016: 2). For the political philosopher Thomas Nail, borders are therefore much more about

In today's control societies, borders have in many ways also become abstract codes, sophisticated algorithms, and digital barriers. However, not only has the form and function of borders changed in control societies but also migration and even having a body has become a different meaning, as the body itself has been transformed into a source of truth in the process of migration and border control. Biometric technologies relocate the border deep into the body, exercising a kind of surveillance from 'within'. Even though it is important not to lose sight of the fact that borders are not democratic, meaning, that not everyone can pass through every border, borders today seem to be nowhere and everywhere at the same time.[18] They can be portable such as ID cards and biometric passports or virtual and thus accessible from everywhere, as it is the case for biometric and genetic databases. This development not only transforms the meaning of migration but also that of bodies in a very material sense.

Against this backdrop, there is a tendency to argue that new biometric technologies are making use of the body as a source of information and an object of control (Dijstelbloem et al. 2011a: 7). While such a perspective regards certain technologies as tools endowed with the power to read, translate, and transform the material body in many ways, the body itself remains plastic matter. What is more, it is argued that the body, as a technologically-readable body, becomes ultimately a component of the machine; however, an entirely disempowered one. In some ways it seems true that "the information revolution was seen as a movement away from the physical, material world [...] but now we can see that the contrast thus created was illusory" (van der Ploeg/Sprenkels 2011: 74), as today it becomes increasingly difficult to separate the material body "from a dataset or an information-processing machine. Our DNA is a code, our medical history is an electronic patient record, our physical vulnerabilities amount to an ICT-generated risk profile and our identity is an algorithmically produced biometric template" (ibid). And yet, it would be misleading to understand the body primarily as a "machine-readable" object, a "source of information" (Dijstelbloem et al. 2011a: 2), or even as a mere "information storage device" (ibid: 12). As a matter of fact, it seems neither very convincing nor particularly helpful to

---

    movement and the redirection of flows of humans, nonhumans, and things, than about physical barriers or static lines on maps.

18 Donald Trump's populist election promise to build a wall along the border to Mexico, and the border fences that have been erected in Central and Southeast Europe in the wake of the so-called migration crisis of 2015, however demonstrate that physical borders are not disappearing but, on contrary, having a revival in some parts of the world.

consider biometric technologies as an autonomous technology or as a "technology out of control" (Dijstelbloem et al. 2011b: 172). Without doubt biometric technologies have politics, as I will demonstrate in the following, but neither do they represent technologies out of control nor is the body a passive object, an open book that can be read and even rewritten without any form of resistance. In what follows, I shall turn to another case to make this argument more lucid.

In January 2010, while helping out as a counselor for refugees in an NGO in Vienna, Cumar, a former client of me who travelled the year before from Somalia through Ethiopia and Libya to Italy, and from there to Austria, reported about the struggle his cousin Kaahin was facing in the UK. After Kaahin's sister and two of his elder brothers had already received asylum in the UK, Kaahin was the last member of the family to escape war-torn Somalia for a better life in Europe. However, by the time Kaahin was applying for asylum in the UK public debates on the abuse of the UK asylum system by refugees[19] were dominating the media and politics. Eventually, these debates culminated in the call for developing "appropriate measures in order to stop this abuse of the UK asylum system" (UK Border Agency 2009b: 3). As someone whose development as a scholar was shaped by the experience as a migrant, I could immediately relate to both Cumar's situation and those debates in different biographical, political, philosophical, and affective ways.

Against the background of the so-called *War on Terror*, post 9/11 security policies, as well as recent global migration movements, new biometrical identification technologies are increasingly used to complement traditional methods of

---

19 According to the 1951 Refugee Convention in Geneva, the term refugee applies to any person who, due to "a well-founded fear of being persecuted for reasons of race, religion, nationality, membership of a particular social group or political opinion, is outside the country of his nationality and is unable or, owing to such fear, is unwilling to avail himself of the protection of that country; or who, not having a nationality and being outside the country of his former habitual residence as a result of such events, is unable or, owing to such fear, is unwilling to return to it." (Convention of 1951, Article 1A (2)) While the definition of asylum seeker varies from country to country, the term usually refers to people who apply for protection as a refugee and are awaiting the determination of their status in the country they applied for asylum. If protection is granted, the person becomes a recognized refugee in that country. The term migrant can have different meanings and often no sharp distinction between migration and flight is tenable or even possible. Take for instance people fleeing life threatening ecological catastrophes such as famine, drought, or toxic environments.

border and identity control.[20] In September 2009,[21] the UK Border Agency[22] announced the launch of its *Human Provenance Pilot Project*. In the wake of debates on the alleged "abuse of the UK asylum system" (UK Border Agency 2009b: 3) and worries about so-called "nationality swapping" – that is, the accusation that refugees would often claim to come from certain war-torn regions such as, for example, Somalia, Iraq, and Syria in order to have better chances for receiving asylum – UK authorities were searching for new methods to "identify a person's true country of origin" (UK Border Agency 2009a).

As a microscopic regime of seeing-knowing-materializing the *Human Provenance Pilot Project* ought to reveal truths about the very essence of material bodies by reading off "ethnic origin" as well as "nationality" (see UK Border Agency 2009a; 2009b) from the bodies of African asylum seekers, using genetic testing and new biometrical identification technologies. Combining genetics and biometrics to verify identity, identity was not only understood as 'ethnic origin' but also as 'national origin'. Despite its name suggests something different, the project did not aim at determining the provenance of human beings as a species, but rather targeted persons who were "making false claims about their

---

20 Here, the question might arise whether the traditional methods are necessarily the less invasive and less repressive ones. For example, it could also be argued that beyond the "epistemological differences between the old techniques of document analysis and interviews on the one hand and DNA analysis on the other, it has to be emphasized that from an ethical standpoint the older techniques are probably no less problematic than DNA tests, because their potential to harm personal privacy is equally enormous, especially when intensive interrogation techniques are used on children. Interviews are not automatically the ethically less challenging technique…" (Weiss 2011: 16). However, rather than theorizing which method of identity control would be the more invasive and repressive one, it seems more reasonable to shift the focus on concrete examples and the socio-economic and political fabric in which particular practices and methods of border control are exercised.

21 Despite its announcement in September 2009, the project was already running at least since February 2009, as a Freedom of Information Act 2000 released on 12 December 2011 states. See UK Home Office 2011.

22 In March 2012, the former UK Home Secretary and current Prime Minster of the UK, Theresa May decided, after the accusation that "officials had abandoned rules", allowing people "into the country without appropriate checks", the splitting up of the UK Border Agency in two separate entities: the UK Border Force, a "law-enforcement body with its own distinctive 'ethos'", and the National Crime Agency (see Mason 2012).

nationality when making their application for asylum" (UK Border Agency 2009a), providing evidence for assessing whether or not applicants were telling the truth about their country of origin. Since the UK Border Agency considered so-called nationality swapping being widely common especially among African asylum seekers, the project solely targeted refugees from Africa, in particular those stating to come from Somalia. Somalia is regarded as a failed state, that is, as a state that cannot guarantee the basic rights of its citizens, therefore, chances for Somali refugees to receive asylum in the UK were high, as long as they could prove that they were indeed fleeing from Somalia.

Departing from the idea that bodies cannot lie, the UK Border Agency's *Human Provenance Pilot Project* operated following the assumption that a combination of DNA ancestry testing, which involved Y-chromosome analysis,[23] mitochondrial DNA analysis,[24] and single-nucleotide polymorphisms (SNP) testing, along with strontium isotope analysis would reveal the asylum applicant's 'true country of origin' (ibid). Applicants were asked to provide a mouth swab and hair and nail samples, which then were tested for DNA and certain isotopes. Although participation was said to be entirely voluntary and only done with the consent of the applicant (ibid), policy documents and manuals for Asylum Screening Unit officers reveal a different picture of the threatened consequences for the applicants.[25]

Mitochondrial DNA tests are used to scrutinize the genetic information stored in mitochondria. Mitochondria are organelles, that is, subunits located in the cells of animals and plants that not only provide energy to body cells by converting food into a form that can be consumed by the cell, but also possess their

---

23 Y line DNA tests look at specific markers on the Y-chromosome that are passed down paternal lines. This was also the reason why the project exclusively targeted male refugees. While sex and gender do not appear in the technical language of the project itself, as embodied categories they were both absent and present at the same time.

24 Mitochondrial DNA tests are used to scrutinize the genetic information stored in mitochondria. Mitochondria are organelles located in the cells of eukaryotic organisms which possess their own, independent genome.

25 The case owner manual, for instance, states: "If an asylum applicant refused to provide samples for the isotope analysis and DNA testing the case owner could draw a negative inference as to the applicant's credibility and if appropriate apply Section 8 of the Asylum and Immigration (Treatment of Claimants, etc.) Act 2004. Section 8 states that where an asylum applicant has behaved in way that is designed or likely to conceal information or mislead the UK Border Agency; it could be seen as damaging the applicant's credibility." (UK Border Agency 2009b: 8)

own, distinct DNA. Y line tests look at specific markers on the Y chromosome that are passed down paternal lines, which was also the reason why females "were unable to be DNA tested using the Y chromosome analysis method because they have two X chromosomes in their cells and not an X and a Y" during the run time of the *Human Provenance Pilot Project,* as the UK Border Agency (2009b: 3) states.[26] Both tests, mitochondrial DNA analysis and Y chromosome analysis, can, under specific circumstances, give information about a person's ancestral heritage since certain genetic variations are more common in certain geographical areas of the world than others. Single Nucleotide Polymorphisms are subtle variations in the genetic code of a person's chromosomes that can also correlate to ethnic origin. However, all these tests are anything but accurate and only provide very limited information (cf. Nature 2009). Isotope analysis is used in archaeology, anthropology, and human geography to date material cultural artifacts, to track historical movements of people, and more recently also to study migration movements of endangered animals as well as environmental influences on them. In the case of the *Human Provenance Pilot Project* it was believed that certain isotopes could determine the "true country of origin of an applicant" (UK Border Agency 2009b: 8) as well as the possible routes the applicant took to get to the United Kingdom.

The UK Border Agency never revealed details about the isotopes under examination to the public. However, the use of skin and nail tissue samples suggests that the tests focused most likely on lighter element isotopes such as strontium, oxygen, and hydrogen (cf. Travis J. 2009). Strontium is a chemical element that belongs to the group of alkaline earth metals and is mostly found in inorganic materials such as rock. Weathering allows the strontium isotopes to trickle into the ground water from where they find their way into plants, animals, and the human body. Similar to calcium, strontium then becomes embedded in bones, hair, and nail tissue. As signatures of isotopes, that is the number of particles in the atomic nucleus, vary according to geographical location and the isotopes incorporated are in constant exchange with the surrounding environment, analyzing the isotopic ratios in nail and hair tissue and matching them against comparison ratios from the country of which the asylum seeker claims to hold

---

26 Although the manual for case owners states that females "can be tested using the mitochondrial analysis method and in the near future it will be possible to test women using SNPS, which is expected to begin during the life of this pilot" (UK Border Agency 2009b: 3), it remains entirely unclear whether or not females were tested during the run-time of the pilot project.

nationality, should have allowed the UK Border Agency to draw conclusions about the place of birth of the applicant.

Originally, the UK Border Agency had planned to use genetic and isotope test results as definitive proof of the applicants' nationality. However, due to heavy critique from the scientific community,[27] the authorities decided not to rely solely on the test results in the evaluation process of individual applicants' cases. After just a year, the project was terminated in March 2010. The planned international review was suspended, the report therefore never published, and the UK Border Agency stated that it has no intentions of continuing the project in the near future without providing any explanation for this decision (see UK Home Office 2011).

The fact that the *Human Proverance Pilot Project* exclusively targeted bodies of color, and in particular Somali bodies, is as little a coincidence as the reason as to why the project was brought to life in the United Kingdom in the first place. The United Kingdom has one of the largest DNA databases in the world with more than five million samples that have been collected claiming their usefulness in fighting global terrorism. DNA here becomes a tool that under the pretense of combating serious crime and terrorism promotes the surveillance and discrimination of large parts of the population, and especially of marginalized social and ethnic groups. In a recent decision the European Court of Human Rights ordered UK authorities to erase from their databases almost one million DNA samples from persons who had never been charged for anything. However, chief constables across the UK have been not only told by the government to ignore the ruling by the European Court of Human Rights but have also been "'strongly advised' that it is 'vitally important' that they resist individual requests based on the Strasbourg ruling to remove DNA profiles from the national database in cases such as wrongful arrest, mistaken identity, or where no crime has been committed" (Travis A. 2009). The fact that the UK's national DNA database is made up predominantly by black males even though the UK Home Office states that actually "black people have lower offending rates than their white counterparts" (MacAttram 2009), as well as the circumstance that by the year 2006 seventy-seven percent of young black males had genetic profiles in the UK

---

27 In an article published in the journal *Nature* (2009: 697), scientists state that "[g]eneticists, and indeed all scientists, should decry the plan and make it clear that the science does not support it [the *Human Provenance Pilot Project*, JB]". In other places geneticists have argued that "mtDNA [mitochondrial DNA, JB] will never have the resolution of specify a country of origin [...] what they [the UK Border Agency, JB] are selling is little better than genetic astrology" (Oates 2009).

national DNA database (cf. Leapman 2006),[28] illustrates that neither the *Human Provenance Pilot Project* nor genetics and biometrics in general could be separated from questions of power and social inequality. This also includes considering the continuities of the UK's colonial policy as a constitutive part of this specific apparatus, given the facts that Somalia was a British colony until 1960 and that the largest Somali community in Europe lives in the United Kingdom (cf. Hopkins 2006). Closely tied to this are economic and political discourses[29] and practices that attempt to restrict migration to the European Union, fortify the EU's borders in the Mediterranean Sea and Northern Africa, gradually transform the (outer) borders into technological ones, and privatize parts of the EU's executive forces. All these practices and discourses represent constitutive material-discursive components of the *Human Provenance Pilot Project* as a techno-apparatus of bodily production from which the question who dies and who lives and at what costs cannot be separated.[30]

---

[28] Similarly, in the US more than sixty percent of all prisoners incarcerated are people of color (cf. Duster 2006) and over forty percent of the genetic profiles stored in national DNA databases were from black people alone, although they make up only thirteen percent of the national population (cf. Greely et al. 2006).

[29] In 2013, David Cameron, former Prime Minister of the United Kingdom, said that he would opt for withdrawing from the European Convention of Human Rights in order to "keep the UK safe" if needed – meaning, in order to be able to deport people who would have "no right to be in the country more easily" from the UK. Following a quite disturbing pragmatism, Cameron also stated, "What we need to do is look and think what is the outcome we want. I'm less interested in which convention we are signed up to" (The Guardian, September 29, 2013). Former Home Secretary and current Prime Minister of the UK, Theresa May went even further, arguing that the plan is to "deport first, and hear the appeal later" – meaning, after the asylum seeker or undocumented migrant has been deported to the supposed country of origin. This would even apply to people who claim to have a right to a family life as article eight of the Human Rights states it. The only exceptions would be cases where the person concerned is threatened by torture and execution after their arrival (cf. Chorley et al. 2013).

[30] According to different estimations between 15.000 and 23.000 undocumented refugees died within the last decade at the borders of the European Union. In the first half of 2015 alone almost two thousand refugees died in the Mediterranean Sea. The only solution for this crisis the European Commission came up so far is the effort to systematically capture and destroy the vessels used by refugees to cross the Mediterranean Sea and to fingerprint all migrants in the near future.

Understanding the UK Border Agency's *Human Provenance Pilot Project* as a techno-apparatus of bodily production means to consider it, far from determining supposed truths about bodies and environments, as an attempt at enacting and naturalizing bodies, boundaries, and territories alike. Functioning as a site where human and nonhuman bodies, technologies, technoscientific and other practices, as well as geologies, politics, and laws, in their entanglements, intra-actively enact the very bodies, boundaries, and territories the project sought to measure, the *Human Provenance Pilot Project* was as much concerned with the goal of producing knowledge (about an imagined essence of bodies) as it was concerned about enacting specific materialities, meaning, material bodies that are eligible for asylum in the UK and those that are not. As I have elaborated in the previous chapter, measurements have material and therefore ontological consequences. As a matter of fact, technologies and technoscientific practices shape, at least to a certain degree, what an object of knowledge is or ought to be. Measurements, in this sense, are not "simply revelatory but performative; they help constitute and are a constitutive part of what is being measured. In other words, measurements are intra-actions [...] material-discursive practices of mattering" as Karen Barad (2012b: 6-7) puts it.[31] Similar to the spirometer that not only produces the entity it is measuring, that is, vital capacity, but also the corresponding bodies, marking bodies of color as less 'vital' by an alleged natural law, also the *Human Provenance Pilot Project* has not so much revealed supposed truths about the bodies analyzed but rather reconfigured them by short-circuiting technologies and technoscientific practices with biology, race, ethnicity, and nationality, determining, but at the same time also materializing, which bodies are eligible to receive asylum in the UK and which are not. The title of an article in the Journal *Biometric Technologies Today* illustrates this idea aptly, stating that biometrics is "not what you know, it's who you are" (Fisher 2008: 7). Genetic and biometric technologies act on the assumption that there is an essence, an immanent truth in material bodies that could be read off technologically. But in fact, it is only through these technologies and practices that the very entities they seek to measure come to matter in both senses of the word. To put it bluntly, there was, and is, no authentic (biological or ethnic) Somali or Kenyan body *before* its enactment through the *Human Provenance Pilot Project*.

Not only the border became a fluid contact zone composed of human and nonhuman bodies, amongst other things, but also the bodies involved embodied the border. In an important sense, the UK Border Agency's *Human Provenance Pilot Project* did not so much mark or represent the border as it performatively

---

31 See also chapter three, in particular page 133 and following pages.

hailed into being both particularly reconfigured bodies and borders. The bodies under investigation themselves became the border, as the isotopes incorporated in hair and nail, the mitochondria in the cells, and the chromosomes of the applicants were linked to particular geologies and territories, determining who belongs to a particular geographical place. Zoē was transformed into bios, bare biological life, such as the isotopes and mitochondria incorporated in the bodies of asylum applicants, became political life.[32] These geomarkers, that is, particular geological and geochemical materials incorporated in bodies that refer to particular places or locations on the planet, not only were sought to link the biological body to political images and imaginations about ethnicity, kin, nationality, the environment, soil, and territory, but also to mark them in a very material sense, essentializing territories, boundaries, and bodies, in the very same movement. Nationality, for the UK Border Agency, became identical with nativity.

Giorgio Agamben reminds us that both the nation and the native share the same Latin root, namely, 'nationem' for 'that which has been born', demonstrating the etymological association of territories and borders with organisms.[33] It has to be understood in this light that for Agamben the nation state,

"means a state that makes nativity or birth [*nascita*] (that is, naked human life) the 19one) of the first three articles of the 1789 Declaration: it is only because this declaration inscribed (in articles 1 and 2) the native element in the heart of any political organization that it can firmly bind (in article 3) the principle of sovereignty to the nation (in conformity with its etymon, *native* [nation] originally meant simply "birth" [*nascita*]). The

---

32 Even though I follow Agamben's differentiation between bios and zoē here, in a certain sense, the very distinction between bare existence, or 'naked life', and political life could also be problematized as thoroughly anthropocentric.

33 The association of the nation with something that is born, something that lives, grows, and can die just like an organism dates back to Greek and Roman political philosophy. Images of the nation state as a quasi-organism and the use of organicist metaphors for describing nations however have especially proliferated in the 19th and early 20th century. The German zoologist, Carl Vogt, for example, drew parallels between the development of organisms and states, proposing the notion of the "state-organism" in the mid-19th century. Similarly, in *The State as a Living Form*, (Staten som lifsform 1916) the Swedish political scientist, Rudolf Kjellén, not only coined the term 'geopolitics' with its five levels that will become an important point of reference for National Socialism – *Reich, Volk, Gesellschaft, Regierung,* and *Haushalt* – but also understood the nation state as a kind of transindividual being, a living organism that precedes both collectives and individuals.

fiction that is implicit here is that *birth* [*nascital*] comes into being immediately as *nation*, so that there may not be any difference between the two moments." (Agamben 2000: 21)

It is this link between nationality and nativity that the *Human Provenance Pilot Project* sought to reinforce; a link that is constantly threatened by migrants, particularly by undocumented ones, "who break the link between nativity and nationality and bring the nation-state system into crisis" (Khosravi 2010: 2).

Even though it is not new to link bodies to particular geological and political territories, until recently the marks were predominantly left on the *surface* of bodies – for example, the color of the skin, the shape of the body, etc. This is not to say that phenotype becomes unimportant but rather that phenomena such as the UK Border Agency's *Human Provenance Pilot Project* aimed at relocating the link between nativity and nationality, between kin and territory, into the very depths of the body. As the project sought to enact not only bodies marked by ethnicity and nationality, but also their corresponding environments, that is, geological and political territories, it aimed at actualizing race and ethnicity as a geopolitical project that is enmeshed with the ideologies of capitalism and the nation state. Rather than charting borders as fixed lines on maps and reading off particular identities as supposed truths from bodies, it was only through the project that naturalized territories and allegedly authentic Somali bodies came into being – bodies that mattered, and subsequently received asylum in the UK, and those that did not matter. Angela Mitropoulos (2016) reminds us that if we think of the border "as a site of invariant bodies moving across a divided but otherwise continuous space—then we begin by assuming, incorrectly, that the properties of bodies (or even 'the subject') and the properties of space can be distinguished". Considering the *Human Provenance Pilot Project* as a techno-apparatus of bodily production suggests that rather than occupying a particular space, bodies and environments were intra-actively constituted as borders themselves. What is more, such an account illustrates that the fact that borders have become ubiquities does not mean that the border is everywhere, but rather that every space-time point can *potentially* become the border for *certain* bodies, which then are not so much "expelled by the border", as they are "forced to *be* the border" (Khosravi 2010: 99).

The *Human Provenance Pilot Project*, as a techno-apparatus of bodily production, did not only translate flows of data into material bodies and vice versa but also the bodies the *Human Provenance Pilot Project* sought to enact were not just any bodies but bodies marked by ethnicity, nationality, and race. The fact that in the case of UK Border Agency's *Human Provenance Pilot Project* exclusively black bodies came into focus makes explicit that some bodies are obviously seen as less trustworthy than others. The black body claiming to come from Somalia itself became a document; one which, it was thought, could not lie and whose true essence could be unveiled by means of certain technologies and technoscientific practices. It is against this backdrop that Donna Haraway's urge not to ignore the potentially dangerous consequences that entanglements of informatics, biology, and politics are bringing with them has to be called to mind. Haraway makes this point especially clear arguing that "lives" are what is "at stake in curious quasi-objects like databases; they structure the informatics of possible worlds, as well as of all-too-real ones" (Haraway 1995b: xix). And indeed, in a certain sense the *Human Provenance Pilot Project* can be seen as contributing to the racist history of the reification of certain human beings constructed as passive, subordinated objects for technoscientific investigations.

Therefore, if biometric technologies are positioned as means that would actually benefit refugees and asylum seekers, as they would afford them "a credible means of establishing their identity, even where they lack other documentation", and if subsequently the conclusion is presented that even though "biometric technology is not free from misuse, [...] biometrics will continue to be an important tool in protecting refugees and asylum seekers" (Farraj 2011: 891, 941), such assessments risk reinforcing not only the turn to the body as a mere passive information storage device but also the idea of race as a plausible biological category. To argue that "biometric technology is [not] inherently racist, but that it is sometimes used to draw and uphold racial lines" (Browne 2012: 77) is true and yet reproduces the belief that there is a distinction between technology on the one hand side and 'its application' on the other. Understanding the case of the *Human Provenance Pilot Project* and the technobiopolitical past and present of the spirometer however demonstrates that such a perspective is an illusion.

CONCLUSIONS: A NEW MATERIALIST THEORY OF TECHNOLOGY AND THE BODY

Donna Haraway reminds us that at the heart of what she terms speculative fabulation, that is, "the practice that studies relations with relations", lies the idea that it "matters what matters we use to think other matters with; it matters what stories we tell to tell other stories with; it matters what knots knot knots, what thoughts think thoughts, what ties tie ties. It matters what stories make worlds, what worlds make stories." (Haraway 2011a)[34] Stories, in an important sense, are not fictions or made up, but rather, as Haraway (1997: 230) emphasizes, "devices to produce certain kinds of meanings" – *and* materialities, one has to add. "Stories and facts do not naturally keep a respectable distance; indeed, they promiscuously cohabit the same very material places." (Haraway 1997: 68)

In a certain sense, both stories above could be read as examples for the power of science and technology over our[35] very lives and bodies. As technologies of control that target the body, both phenomena could serve as examples par excellence for narratives that operate through rhetorics of loss, emphasizing the technoscientific colonization, disciplining, and objectification of material bodies. However, both stories could also be read differently, namely as technoscientific attempts at silencing the forces and flows of material bodies that, in doing so, have not functioned as intended.

Throughout this book I have argued that it seems epistemologically and politically limiting to assume that technologies are first and foremost congealed power relations, facing a mute and passive body. Emphasizing the limited rather than the pervasive power of technologies as well as raising the question for the potentials of material bodies to be unruly and to resist the attempts of their technological and technoscientific access does not mean to reinstate a dualism between technologies and bodies once again. Here, it is crucial to differentiate between *technology* and *technologies*. While *technology* denotes a particular mode of being and knowing, and precisely therefore cannot be regarded as the Other of the body, but, on the contrary, has to be understood as always a part of 'us', *technologies* refer to relational matrices, to specific worldly constellations, to entangled multiplicities. Therefore, *technologies* are never mere 'technical', but always political, economic, cultural, and social. What is more, they are, to put it with Bruno Latour, always 'full of humans and nonhumans'. The concept of the techno-apparatus of bodily production as a speculative lens and a site is what allows the taking into account of these processes of becoming-with-technologies. The notion of becoming-with-technologies does not refer to a co-production of bodies and technologies but rather to specific material reconfigurings. It is also for this reason that even though technology is not the Other of the body, technologies and bodies in their entangled becomings and reconfigurings

---

34   See also Haraway (2011b).

35   Importantly, however, the claim that the *Human Provenance Pilot Project* targeted 'our' lives and bodies has to be already questioned for the reason that it targeted not *any* bodies but primarily bodies marked by race and ethnicity.

are not essentially the same. This is precisely the nature of entanglements, as Karen Barad reminds us: "One is too few, two are too many" (Barad 2014: 178).

Against this backdrop, rejecting both the picture of the body as a mere object upon that powerful technologies act as well as the idea of technologies as mere artifacts or systems that would only *interact* with our bodies opens up an avenue that promises to disrupt prevalent understandings of technology and the body, and how they relate to one another. If power is something that emerges out of the entanglements of humans and nonhumans, as I have outlined in detail in the previous chapters, rather than something that can be acquired and possessed by individuals, it follows that the assumption that material bodies are mere passive objects seems questionable the very least. Analyzing the biopolitical past and present of the spirometer and the case of the UK Border Agency's *Human Provenance Pilot Project*, in this light, suggests that material bodies are not only inseparable from the techno-apparatuses of bodily production through which they come into being but also that they have to be considered as active parts of those apparatuses.

Considering the UK Border Agency's *Human Provenance Pilot Project* as a techno-apparatus of bodily production allows us to understand how bodies marked by race and ethnicity, among other categories, come to matter in their entanglement with technologies and technoscientific practices, but also geologies and political discourses. Instead of treating material bodies as mere passive moldable matter, as not much more than a kind of information storage device that can be accessed through the means of genetic and biometric technologies, such an account provides us with the possibility to think material bodies (as parts of apparatuses) as agentic entities. After all, such an analysis suggest that attempts at reading off 'ethnic origin' and nationality from the bodies of African asylum seekers failed, and it would be too easy to see this as an outcome of the circumstance that the *Human Provenance Pilot Project* was scientifically flawed or that the technologies applied were not accurate enough. Instead, the perspective I proposed in this book puts to the fore the idea that neither did the applied technologies function according to the interests inscribed into them nor did the bodies concerned played along as expected; meaning, neither the bodies involved could been fixated on mere passive objects of knowledge, nor were the isotopes under investigation incorporated in the ways expected. Even if these attempts were successful, nothing less than the fact would have been ignored that while it is true that isotope signatures might vary according to geographical location, this does not mean that isotopes would necessarily stick to national borders. The idea that the isotopes on the Somali side of the border are fundamentally different from those on the Kenyan side lacks plausibility. As for the genetic examination of the applicants: Even though it seems to have come as a surprise for the UK Border Agency, it might not be a surprise at all to realize that genes are not aware of the political concept of nationality. Even if the tests had been able to provide exact information about the ancestry of the applicants, determining nationality as a political and legal concept is something entirely different. Equating the place of birth with nationality of an applicant is not only deeply flawed, but also ignores that people move and do not necessarily accept the fact that they were born at a particular place as a matter of fate. Neither nationality nor cultural identity are naturally given attributes that can be pulled over human beings, nor are cultures identical with nation states in the sense of containers with defined boundaries in which individuals with the same beliefs, ideas, and practices can be put into. Rather, it makes more sense to regard cultures as always fluid, gaseous, or plasmic formations without clearly defined borders. What is more, within such a framework individuals always simultaneously belong to a multitude of cultures because the term culture does not function any longer as an euphemism for nationality, ethnicity, or religion (cf. also Singer 2012).

It is in this sense that theorizing the *Human Provenance Pilot Project* as a techno-apparatus of bodily production demonstrates, on the one hand, that biometric identification technologies such as DNA and isotope testing represent new technologies of surveillance, especially in the process of migration, raising serious political and ethical questions; not only because these technologies and technoscientific practices tend to reessentialize and rebiologize race (for example, by equating race, phenotype, and genotype), but also because they constrain kinship and family to biology.[36] What is more, the project, as an integral part of the UK's border machine, demonstrates that the border neither begins nor ends at the geographical or political borders of the EU, but is relocated into the depths of the body. On the other hand, however, the analysis above also suggests that even a powerful, repressive, and anti-democratic technology of control such as the *Human Provenance Pilot Project*, in the end, was not powerful enough to mute the forces and flows of material bodies (as parts of the apparatus); nor appears it to be the case that certain political interests and beliefs could be inscribed into the technologies used with the intended effects.

---

36 That being said, in other cases it could also be argued that it is precisely genetics and reproductive technologies that may undermine the idea of parenthood as a natural relationship that would ground in biology, and that genomics also possess the power to "de-biologise the family on the grounds of scientific insights" (Weiss 2011: 17).

The same applies for the spirometer. With the practice of systematically lowering the measured values in vitality for people of color, the spirometer reproduces the idea of a racially defined and scientifically definable deficiency that people of color, and black people in particular, would have. In doing so, the spirometer reproduces the pathologization of black bodies as well as the idea of a biologically flawed black body. And yet, similar to the UK Border Agency's *Human Provenance Pilot Project*, considering the spirometer as a techno-apparatus of bodily production provides a different perspective, one which, rather than overemphasizing the omnipotence of science and technology, suggests that if it was the intended socio-political function of the spirometer to prove in the wake of slavery and forced labor in the US of the second half of the 19th century that black people were not made for a life in freedom because of their alleged lesser efficiency of their organs, it ultimately failed. Likewise, if it was the function of the spirometer to discriminate black workers who had fallen ill in the late 20th century, preventing them to receive compensation payments, it also ultimately failed. In the lawsuit against *Owens Corning*, the court decided "after hearing testimony from medical experts [...] that there was no reason to use different medical standards for whites and blacks" (Los Angeles Times: 1999). Relying on the power of the spirometer as a tool for discriminating black workers hence proved to be unsuccessful.

Once again, this is not to deny the usefulness of the spirometer in detecting respiratory diseases, nor the discriminatory effects this instrument had on the lives of so many people, but on the contrary to open up a different perspective that might allow us to see a fuller picture. Above all, such an account suggests that it is too easy to regard technical objects as condensed and congealed social power relations that function precisely as intended. Furthermore, it could be argued that rather than 'objectively proving' that vital capacity, vitality, and the efficiency of organs would differ according to race by 'natural law', the spirometer demonstrates the exact opposite, namely that vital capacity and with it also material bodies proved to be resistant against attempts of their standardization, categorization, and normalization. From such a perspective, it could even be argued that the dynamic nature of bodies and bodily organs disturbed, if not disrupted, both the belief in innate differences which would render black bodies less vital compared to white ones by 'natural law', and the idea that technologies, understood as condensed social power relations, would function precisely according to particular interests that have been inscribed into them.

As a matter of fact, Lundy Braun emphasizes that for the spirometer to function as intended, reliable standards for measuring vital capacity had to be developed. This, however, was anything but easy since, "many factors that might influence the assessment of lung function [had to be taken into account, JB]—the technicalities of the instruments, operation methods, patient behavior, physical activity, statistical methodology, and anthropometric variables. Despite such careful analysis, lung capacity, the entity measured, proved more resistant to standardization than anticipated." (Braun 2014a: 110-111)

The fact that the bodily flows and forces "had to be 'tamed'", as Braun (2014a: 113) emphasizes, in order to be able to categorize lungs as 'normal' or 'abnormal', demonstrates that bodies and bodily organs cannot be regarded as mere passive objects, but rather proved to be resistant against attempts of their standardization, categorization, and normalization. It is precisely for this reason that the belief in the power of science and technology, or in the spirometer in this particular case, to prove difference in the efficiency of organs and bodies according to race, and consequently the different worth of human bodies, ignores the dynamic nature of material bodies. The body and its functions are not constant and static but subject to continuous changes. This fact, in turn, gets in the way of a classification and categorization of bodies as well as of the attempts at proving technoscientifically that bodily capacities and functions would differ according to race by 'natural law'. It is against this backdrop that it could be said that the spirometer with its so-called 'race correction' function embedded in its hard- and software, ultimately, "reinforces—and buries—the idea of 'naturally occurring' differences in lung function" (Braun 2014a: xix), and consequently the idea that lung function or vital capacity is directly related to race and ethnicity.[37]

Correlation does not imply causation. Hence, even if it were true that there are differences in lung capacity and vitality between 'ethnic' groups or even according to race, the question remains just how much these differences have to be regarded as innate or, on the contrary, have to be seen as effects of environmental and socio-economic influences such as air pollution, poverty, little or no access to health care and education, and unhealthy working conditions, to name

---

37 Lung capacity is not the only entity in science and medicine related to the body or parts of it that has been racialized. Anne Fausto-Sterling traces and debunks "the belief in a racial hierarchy of bone strength", and in doing so points out that bone strength is not a "race trait" but rather "individual bone structure emerges from a combination of genetic possibility, diet, exposure to sunlight, skeletal biomechanics, forms of exercise and physical labor, plus other contributors that we have yet to articulate with any clarity" (Fausto-Sterling 2008: 659).

but a few. Epigenetic studies[38] demonstrate that the environment can affect modifications in the structure and function of cells influencing which genes are given expression and consequently which proteins develop. These changes in the cellular environment, also called epigenetic markers, "can persist and sometimes be passed from parent to offspring" (Guthman/Becky 2012: 491), affecting, yet not determining, protein synthesis and subsequently not only the function of organs but even phenotype. Differences in lung function could therefore also very well be regarded as the embodied history of imperialism, colonialism, classism, racism, and social inequality. In fact, such an argument is not a particularly new one as already in the mid-19th century black intellectuals such as Frederick Douglas, Kelly Miller, and later also W. E. B. Du Bois put forward the influence of social and economic conditions on black bodies.

In his speech at Western Reserve College in the year 1854, Frederick Douglass, a former slave and abolitionist intellectual, argued for a universal understanding of human rights. For Douglass, a "diverse origin does not disprove a common nature, nor does it disprove a united destiny. The essential characteristics of humanity are everywhere the same." (Douglas 2000 [1854]: 296) In doing so, Douglass contested the belief that human beings would differ in quality according to race, ethnicity, or nationality. Referring to the emerging biological racism against Irish people in England and the US, depicting them, and in particular so-called Irish Travellers, as uncivilized ape- or gnome-like creatures, and

---

38 Epigenetics studies the mechanisms that affect how genes are expressed or suppressed, and how that in turn affects protein synthesis. The term epigenetics was coined in the early 1940s by the embryologist Conrad Waddington who described how the environment influences gene expressions and subsequently the development of organisms. Instead of understanding genes as a kind of 'blueprint', epigenetics positions itself against the idea of genetic determinism. Hence, organisms are understood as self-organizing systems that are 'open' to environmental influences. The body and its 'parts' are characterized through a fundamental plasticity. As an approach emphasizing the power of the environment to leave marks on the genome and the body, epigenetics, however, may reinforce certain discourses centering on questions of care (for oneself) and consequently may also evoke stereotypical images; such as, for example, the idea that pregnant women, as the 'initial environment' of the unborn child, not only have to discipline themselves according to specific medical orders and moral beliefs but also have to have themselves monitored constantly, with the effect of creating feelings of concern and anxiety as well as of responsibility for events and processes that might not even be in the hands of the individuals concerned (cf. Müller/Kenny 2014).

hence as an inferior race, Douglass challenged racial traits as inherent by foregrounding the influence of socio-economic variables on the bodies and lives of racialized people.

"I say, with no wish to wound the feelings of any Irishman, that these people lacked only a black skin and woolly hair, to complete their likeness to the plantation negro. [...] The Irishman educated, is a model gentleman ; the Irishman ignorant and degraded, compares in form and feature, with the negro ! I am stating facts. If you go into Southern Indiana, you will see what climate and habit can do, even in one generation. The man may have come from New England, but his hard features, sallow complexion, have left little of New England on his brow." (Douglass 2000 [1854]: 294-295)

Similarly, the mathematician, Kelly Miller explicitly contested Hoffman's idea that people of color, and in particular black people, in the Reconstruction era of the US, that is, after the abolition of slavery, would deteriorate "physically and morally in such manner as to point to ulterior extinction, and that this decline is due to 'race traits' rather than to conditions and circumstances of life" (Miller 1897: 2). In a scientifically eloquent way, Miller unmasks Hoffman's explicit racist bias. Collecting and interpreting dozens of statistics, he shows that socio-economic conditions and social inequality is the reason for a rising death rate in black people, and not some hidden biological causes. As a mathematician, Miller immediately realized Hoffman's lack of scientific rigor hiding behind his racist beliefs.

"The data are so slender and the arguments are so evidently shaped to a theory, that we are neither enlightened by the one nor convinced by the other. But the author's judgment must be justified. The gloomy warning comes with Catonian regularity at the end of each chapter. Listen to his last words: 'A combination of these traits and tendencies must in the end cause the extinction of the race.' If the Negro is inferior in vital function and power to the Caucasian, he will be a public benefactor who scientifically demonstrates the fact. But the colored race most stubbornly refuses to be argued out of existence on an insufficient induction of data and unwarranted conclusions deduced therefrom." (Miller 1897: 19-20)

Just a year later, the sociologist, W. E. B. Du Bois not only problematized the practice of drawing conclusions about the health of all black people by referring to a handful case studies conducted in large cities, because in doing so "the physical condition of all these millions, is made to rest on the measurement of fifteen black boys in a New York" (Du Bois 1898: 14), but also highlighted that many statisticians of his time (including Frederick Hoffman) had failed to stratify their findings by social and economic status. A more rigorous look at statistical data would have shown "that black mortality rates differed enormously based on environmental factors" but also "that the health outcomes of African Americans were entirely comparable to those of immigrant groups with similar economic resources" (Wolff 2006: 85), in particular Irish, Italian, and Eastern European immigrants. Since both economic status and (as a direct consequence of it) health and illness can be inherited, it seems plausible that members of lower classes, and therefore also many people of color who had been oppressed for so long, may have lower vital functions compared to whites.

Therefore, without ignoring its potentially fatal effects, it would be a mistake to understand the spirometer as mere politics by others means or as an instrument that functions precisely according to specific (individual or collective) interests that have been inscribed into it, categorizing, disciplining, and subjugating (certain) bodies. In fact, it is the idea that technologies merely represent politics by other means that leads to an all too hasty overrating of the powers and effects of science and technology. In this regard, many critical approaches on the one side and technocratic and scientistic approaches on the other, ironically, seem to share a similar assessment of the omnipotence of science and technology. In many cases this is precisely the stuff of which "scare stories" (Hacking 1982) are made, which are all too often paralyzing instead of empowering. In fact, many large scale and as immensely powerful described technologies that aim at the control, regulation, and optimization of the lives of the population in every aspect, from birth to reproduction to death, through the means of statistics, scientific theories, and particular technologies have often failed in achieving the "intended effect" (Hacking 1982: 289). This is also true even for the most repressive and monstrous technologies and technoscientific practices, that is, eugenics, racial hygiene, and forced sterilizations. Once again, it is not to ignore the devastating effects these technologies and the technoscientific practices involved had on the bodies and the lives of many people – in particular, on women, people of color, and those living at the margins.[39] Rather, along the lines of

---

39 The question of reproductive freedom, for example, cannot be separated from questions of social inequality. In the US poor women and women of color, in particular, are not only more likely to have undergone sterilization in contrast to white women, but they also have higher rates of miscarriage. The question of technology here, becomes not only a question of what effects certain technologies have on the lives and bodies of women but also the question of who has access to these technologies and who does not, who speaks for whom, and from which position, what counts as a choice, for whom, and at what cost?

Foucault's view that power is not possessed but rather exercised and that where there is power there is also the *possibility* of resistance, it seems too easy to believe that technologies are first and foremost materialized power relations that function as intended and are therefore fully calculable.

Theorizing the biopolitical past and present of the spirometer and the case of the *Human Provenance Pilot Project* as techno-apparatuses of bodily production makes explicit that in both events technologies and politics indeed cannot be separated from each other, that it is misleading even to see them "as *interacting*: there is no categorical distinction to be made between the two", as Donald MacKenzie (1990: 412-413; italics JB) puts it. In a similar and yet different way to approaches that understand technologies as technical artifacts and systems, believing that certain individual or collective interests and power relations could be inscribed into them, and that there would be almost a kind of guarantee that technologies function precisely according to the interests and programs inscribed into them, a new materialist theory of techno-apparatuses of bodily production demonstrates that technologies can neither be reduced to mere artifacts or systems, nor is it the case that technologies function precisely as intended. As I have argued before, it appears considerably more promising to understand technologies as relations, as processes of material reconfigurings, or as events in a more Deleuzo-Guattarian sense, instead of as certain objects with clear and definable properties and boundaries.[40] From such a perspective it makes little sense to look at technical objects cut off from their entanglements. Instead, the apparatus as a

---

40 For this reason it makes also little sense to define what technology *is*, for it seems hardly possible to define what the specific *technical* in technologies would be. In fact, "no general definition of technology [is] possible", as Val Dusek (2006: 36) states. This becomes, for example, obvious if one turns to the question why it seems so clear to many of us to consider, say, a wheelchair as a technology but an 'ordinary' chair only as furniture? Why is an elevator clearly technological but stairs not? Why would nobody deny that a typewriter and perhaps even more a computer are technologies whereas pen and paper are not? And what would be the crucial difference between a mere tool and a technology? As a matter of fact, such questions are not only difficult to answer but also seem to miss the point. Therefore, a relational understanding of technology outlined in this book makes explicit that a clear-cut distinction between technology and technique is only hard to maintain because technology also always comprise of specific knowledges and practices. While technologies as I have argued in the previous chapters drawing on the work of Bruno Latour and Donna Haraway, refers to processes of mutual mobilization and reconfigurings, a relational matrix of humans and nonhumans in their entanglement with one another.

whole[41] has to be taken into account, that is, the forces and elements that constitute it, in order to be able to understand how particular bodies, technologies, and technoscientific practices are entangled with questions of power. For example, in the case of the spirometer and the *Human Provenance Pilot Project*, politics, economics, racial discourses and images, nationalism, medical perspectives on the body, beliefs in the power of science and technology as neutral instruments for revealing truths about the supposed nature of material bodies, and much more marked constitutive intra-acting material-discursive components. Thus, it is not enough to argue that politics becomes embedded into technologies through design decisions or the appropriation of technologies by different social groups; and even less is it enough to stress that technologies have to be regarded first and foremost as congealed social power relations. A "technical element remains abstract, entirely undetermined, as long as one does not relate it to an *assemblage* it presupposes", as Deleuze and Guattari (2004: 439) write. Only by taking concrete techno-apparatuses of bodily production into account it becomes possible to see what technologies and bodies can do, how they materially reconfigure one another and with what consequences. By taking in such a perspective it becomes apparent that even though technology is not the Other of the body, and that even though technologies cannot be reduced on mere politics by other means, technologies do shape the ways of living and dying together (cf. also Haraway 2008). This is precisely why it is necessary to differentiate between technology as a mode of being and knowing, and technologies as relational matrices or specific worldly entanglements.

Applying an understanding of the techno-apparatus of bodily production as a speculative tool for investigations into power – both restrictive power (*potestas*) and enabling, affirmative power (*potentia*) – to the spirometer and the *Human Provenance Pilot Project*, not only it becomes evident that technologies have always politics, but also that there is no given, that is to say, no a priori, boundary between technologies and material bodies. Donna Haraway (1991a) stresses

---

41 As I have outlined in the previous chapter, techno-apparatuses of bodily production are not things in the sense of actual objects with inherent properties somewhere 'out there' but open-ended material-discursive practices and relations, and therefore have no pre-determined borders. However, exploring the spirometer and the *Human Provenance Pilot Project* as techno-apparatuses of bodily production demonstrates that not every component of a given apparatus matters the same way. Thus, the question has to be what matters, in which ways does it matter, with what consequences, for whom, and what is excluded from mattering.

that our bodies do not end at our skin but are always entangled with a multitude of other human and nonhuman bodies around and even within us. In this light, theorizing the spirometer as a techno-apparatus of bodily production reworks 'vital capacity' as the product of bodily organs and functions, the 'environment', socio-economic conditions, technical instruments, as well as medical and scientific theories and practices. On the other hand, analyzing the *Human Provenance Pilot Project* makes explicit that human bodies are not closed systems, and never fully human as they are always comprised of other bodies, organic and inorganic ones alike, such as the mitochondria within our cells or the isotopes incorporated into organ tissue through metabolism. Therefore, both phenomena disrupt the belief in preexisting determinable clear-cut borders between the body and technology, the human and the nonhuman, inside and outside, or a system and its environment. Where does the human body end and where does the nonhuman begin, if one looks at isotopes? Where is the border between the material and the discursive, the factual and the textual, when looking at genes? Genes are neither discrete objects (or even things-in-themselves), nor are they mere social constructions. "A gene is not a thing, much less a 'master molecule' or a self-contained code", but rather signifies "a node of durable action where many actors, human and nonhuman, meet", as Donna Haraway (1997: 142) puts it. Genes are thus itself material-semiotic assemblages. The fact that the number of microbes in our bodies outnumbers human bodily cells ten to one and that the specificity that marks the human genome can only be found in about ten percent of the cells that occupy the commonplace called 'the human body'[42], while ninety percent of the body cells are packed "with the genomes of bacteria, fungi, protists, and viral residues" which are "necessary for our survival" (Åsberg 2008: 266), highlights that it is anything but easy to answer what constitutes the 'humanness' of the human body. Already such a rather simple view on the molecular level demonstrates that bodies always depend on their relations to other bodies and cannot exist independently, that is, cut loose from them. The human body is always already a more-than-human body, or, as Donna Haraway (2003b) puts it, a multitude of "companion species". Looking at isotopes, genes, and many other phenomena, the question as to where to locate the dividing line between inside and outside, the body and 'its environment', becomes entirely unclear.

---

42 See, for example, *The Human Microbiome Project*, an initiative of the US National Institutes of Health that ran from 2008 to 2013, aiming at identifying all microorganisms that are found in association with healthy and diseased humans.

If bodies do not have defined boundaries, if their lines are always fluid rather than solid, does it then even make sense to talk of the body as an object that possesses an inherent essence? Or would it not make more sense to understand bodies as temporary, always-shifting material effects of intra-acting biological, political, economic, technological, and other forces? In *Nietzsche and Philosophy*, Deleuze (1983: 40) highlights that bodies are "constituted" by "relationships of forces". These forces, that can be active or reactive, are what lie behind the ceaseless reconfiguring, the ongoing becoming, of material bodies in their entanglements with other bodies and things. Even though "[t]here is will to power in the reactive or dominated force as well as in the active or dominant force" (Deleuze 1983: 53), force is not to be confused with power, because a force "is what can, will to power is what wills (*La force est ce qui peut, la volonte de puissance est ce qui veut*)" (ibid: 50). While not separable from them, the will to power is never "identical" with forces.

"The will to power is, indeed, never separable from particular determined forces, from their quantities, qualities and directions [...] Inseparable does not mean identical. The will to power cannot be separated from force without falling into metaphysical abstraction. But to confuse force and will is even more risky. Force is no longer understood as force and one falls back into mechanism..." (Deleuze 1983: 50)

Since for Deleuze, forces are dynamic, it cannot be anticipated in advance what they can do. Consequently, it makes little sense trying to define what the essence of a force is without losing oneself in metaphysical nit-picking. However, there is also no need for such a move. I agree with Bruno Latour that philosophical investigations have to "start with the assumption that everything is involved in a relation of forces but that I have no idea at all of precisely what a force is" (Latour 1988: 7), so that there is no "need [for] a-priori ideas about what makes a force, for it comes in all shapes and sizes" (ibid: 154). What, however, is important are the entanglements, the phenomena they enact, and the consequences that are going along with them. It is in this sense that a new materialist theory of techno-apparatuses of bodily production allows for an understanding of bodies as always shifting fields of forces,[43] as relational phenomena to emerge. From

---

43 The term 'field' is not to be understood in a topological sense here, but as referring to relations and entanglements. A body in this sense is more of a temporarily effect of particular relations and entanglements of intra-acting forces, more a "figure" than a "structure" (Deleuze 2003: 20).

such a point of view, bodies exist "quantum-like in multiple states", as Sandy Stone (2002: 524) put it for the body of the mestiza.

As bodies cannot be separated from the very techno-apparatuses of bodily production through which they come to matter, what a body can do is relative to, but not determined by, the very techno-apparatus through which it comes to matter as a specifically re(con)figured body. The *Human Provenance Pilot Project*, for example, represents a techno-apparatus that aimed at producing bodies with specific attributes such as ethnicity, nationality, and race, but also sex and gender. Alongside the bodies, which the *Human Provenance Pilot Project* sought to materialize, also specific attributes of these bodies and specific legal, political, and ethical effects were sought to be enacted – such as being a legitimate asylum seeker who is eligible for receiving asylum in the UK, or being an illegalized body, someone who would only pretend to be a refugee and thus would have no right to receive asylum. Similarly, in the case of the asbestos workers the spirometer as a techno-apparatus of bodily production was employed not only to enact specifically re(con)figured bodies, that is, bodies with a low vital capacity by 'natural law', but subsequently also bodies that were not eligible for compensation in the case of disability and illness. Both the case of the spirometer and the *Human Provenance Pilot Project* aptly demonstrate that one of the most pressing issues of a political philosophy of technology today lies in distinguishing the entanglements that aim at more liveable worlds from those that abandon and kill.[44] It is precisely for this reason that departing from the idea that technology is not the Other of the body does not mean to theorize technologies ahistorically and apolitically. On the contrary, understanding technologies as relations, as processes of reconfigurings, urges us to shifts the focus on the ethical and political consequences that are going along with them.

"We will never find the sense of something (of a human, a biological or even a physical phenomenon)," Gilles Deleuze writes, "if we do not know the force which appropriates the thing, which exploits it, which takes possession of it or is expressed in it. A phenomenon is not an appearance or even an apparition but a sign, a symptom which finds its meaning in an existing force." (Deleuze 1983: 3)[45] However, because forces "are never determining, even when apparatuses are

---

44 See also page 91.
45 In *A Thousand Plateaus* Deleuze and Guattari (2004: 257) actualize this idea, saying that we do not know anything "about a body until we know what it can do, in other words, what its affects are, how they can or cannot enter into composition with other affects, with the affects of another body, either to destroy that body or to be destroyed

reinforcing" (Barad 2012c: 54), it cannot be anticipated in advance what a body can do.[46] The analytical focus has to be on the apparatuses through which specifically re(con)figured bodies materialize. But then what does it mean to regard material bodies as agentic; particularly, if the body cannot be understood as a distinct entity with a priori and transcendent boundaries and properties?

As put it in the previous chapters, agency is not a property of discrete subjects or objects. John Law brings up the "analytical question" whether an agent "is an agent primarily because he or she inhabits a body that carries knowledges, skills, values, and all the rest? Or is an agent an agent because he or she inhabits a set of elements (including, of course, a body) that stretches out into the network of materials, somatic and otherwise, that surrounds each body?" (Law 1992: 384.) In the latter sense, it is not the body that is agentic, since there is no such thing as the body as a discrete, self-contained entity. Instead, agency has to be regarded as an effect of networks, or better of entanglements of humans and nonhumans. Agency then is the capacity of entangled forces and entities to act as quasi-agents with trajectories of their own; while 'to act' here means to make a difference in worldly becomings. Such a relational or distributive understanding of agency not only allows the taking into account of material bodies as agentic but also promises richer analyses of contemporary technobiopolitical entanglements with regard to the question of how matter and meaning simultaneously materialize through specific techno-apparatuses of bodily production, and along with which consequences. These processes of materialization or becoming-with are not to be understood as willful acts. There is no individual or collective agent that has an overview about what is going on, calling the shots. Rather, processes of materialization have to be understood as a dynamic play of intra-actively entangled material-discursive forces, materializing particular phenomena and stabilizing them, at least for a certain period of time, while rendering others impossible. Techno-apparatuses of bodily production and the bodies they materialize are consequently always embodied, meaning, spatio-temporally situated bodies. The materialization of bodies is indeed an effect of power (since power acts on forces as mentioned above) but it is not only an effect of *social* power relations and discourses, nor is power distributed symmetrically, meaning equally, within techno-apparatuses of bodily production.

As in both cases discussed race takes in a central function, the question arises what a new materialist account that acknowledges the body as a material-

---

by it, either to exchange actions and passions with it or to join with it in composing a more powerful body."

46 The emphasis here lies on *a* body not *the* body.

semiotic actor, that is, a generative axis of the techno-apparatus of bodily production, as Donna Haraway (1991a) states, entails. What does it mean to argue that objects and, for that matter, also bodies neither pre-exist as such, nor that they can be understood sufficiently as mere socio-historical constructions or discursive effects? Does such an account circumvent the limitations of social constructionist theorizations of race and allow for a deeper understanding of the very processes through which race comes to matter in the context of the technosciences – not only as meaning and ideology but also as racialized matter and bodies? And what might such an account set against attempts at a reterritorialization[47] and reessentialization of race? Engaging with race, and even more with the question of the materiality of race, is never innocent. It took a very long time for scholars of color as well as other critical thinkers and political activists to deconstruct the belief that race is a naturally occurring (biological) attribute of the human species. However, does this mean that race and racialized bodies are mere socio-cultural and linguistic constructions?

The Deleuzian philosopher Arun Saldanha (2006) argues that race has been considered for too long as a problem of representation within the humanities and the social sciences. For Saldanha, constructionist and poststructuralist accounts of materialization with their focus on the question 'how race is known' fail "to understand processes of racialization", simply because race is not only a problem of how 'we' know race or "how people think about skin colour" (Saldanha 2006: 22). While performative understandings of the processes of materialization of bodies state that there is a materiality to signs, they would refuse to extend this idea to material bodies. That being so, the body and its 'inside', the bones, blood, organs, genes, hormones, and so on, would ironically still remain a constitutive outside. Saldanha thus concludes that,

"a Butlerian critique can rightly question the 'naturalness' of a bedrock of phenotype posited by, and justifying, racial discourse. But such critique halts abruptly at the deep gorge between racist discourse (which it attacks) and phenotypical matter (about which it will not say anything). Is not phenotype itself shaped by cultural practice? Does phenotype ever resist its 'performance'? By not allowing anything from across the gorge to enter her

---

47 I use the notion of reterritorialization here for describing processes of reinforcing and renaturalizing race. Moreover, the term reterritorialization denotes attempts of reestablishing places and territories but also alleged essences, truths, and social power relations that had been contested and, more or less successfully, deconstructed by scholars of color and other critical theorists and activists.

critique, Butler ultimately remains complicit with what she attacks: the metaphysical positing of an inert exteriority to language." (Saldanha 2006: 12)

Contesting Butler, Saldanha puts forward the need for a materialist theory of race in which material bodies themselves "play an active part in the event called race" (Saldanha 2006: 9). Borrowing Michel Serres's notion of "viscosity", Saldanha argues in what follows that race would not be so much "an arbitrary classification system imposed *upon* bodies", but rather an effect of the many ways "bodies themselves interact with each other and their physical environment. The spatiality of race is not one of grids or self/other dialectics, but one of viscosity, bodies gradually becoming sticky and clustering into aggregates." (ibid: 10) Taking in such a perspective, it would not only become possible to know "what race really is, that is, what it can be" but also the "molecular energies of race [could] be sensed, understood, and harnessed to crumble the systematic violence currently keeping bodies in place" (ibid: 22-23). What is more, such an affirmative understanding of race would demonstrate that race does not "need to be about order and oppression, it can be wild, far-from-equilibrium, liberatory" (ibid: 21).

Having in mind that "race is the kind of category about which no one is neutral" (Haraway 1997: 213), it seems indeed shortsighted to believe that racialized bodies are mere linguistic constructions with no material effects whatsoever. But does asking for what race "really *is*" (Saldanha 2006: 22), in the end, not go against attempts to understand race as an eclectic, contradictory, and always-already 'hybrid concept' (in Michel Serres's term)? Frankly, even though Saldanha offers a bold account of how racialized bodies come to matter beyond essentialism and linguistic constructionism, it remains unclear what exactly his affirmative plea for a molecularization of race, that is, for breaking up race "into a thousand tiny races" (Saldanha obviously follows Deleuze and Guattari's call for arriving at "thousand tiny sexes" here),[48] can contribute to anti-racist politics; especially at a time when the re-biologization and re-racialization of certain bodies is again gaining ground in genomics, molecular biology, and evolutionary theory.[49]

---

48 Cf. Deleuze and Guattari (2004: 235); cf. also Elisabeth Grosz's (1993) "A Thousand Tiny Sexes: Feminism and Rhizomatics".
49 A recent example for this is Nicholas Wade's "A Troublesome Inheritance: Genes, Race and Human History" (2014), arguing that natural selection would have led to differences between groups of human beings with regard to intelligence and socio-cultural development. For Wade, the socio-political effects of this could be witnessed

My aim here is not to develop a theory or even a philosophy of race. That being said, I do nevertheless believe that a re(con)figured concept of the techno-apparatus of bodily production, as outlined throughout this book, might be able to provide us with an understanding of not only how racialized bodies come to matter through technoscientific (and other) practices in both senses of the word – that is, how they materialize and become meaningful in the very same movement – but also an understanding of race as a relation *and* a technology; racism, then, would denote the violent system and practice of producing and legitimating social inequality.

In the special edition of the journal *Camera Obscura*, entitled "Race and/as Technology", Wendy Chun (2009) and Beth Coleman (2009) put forward a similar understanding of race. Chun states that understanding race as "a technology" would shift "the focus from the *what* of race to the *how* of race, from *knowing* race to *doing* race by emphasizing the similarities between race and technology" (Chun 2009: 8). As "a disruptive technology", race would change "the terms of engagement with an all-too familiar system of representation and power" (Coleman 178). At the same time, for Coleman, race would also stay for "'algorithms' inherited from the age of Enlightenment" (Coleman 2009: 184). Since algorithms once had to be programed, such an understanding suggests that race could also be reprogramed. Despite the fact that such a perspective contests prevalent understandings of race as either a transcendental, ahistorical truth or as a mere linguistic construction, it is also confronted with the problem that not only the phrase "re/programing" has the overtone of a willful act (maybe even done by the autonomous subject), but also once again nonhumans (that is, technical instruments, mitochondria, chemicals 'within' and 'outside' of the body, and much more) and the roles they might play within the processes of the materialization of race, remain in the shadows.

In contrast to this, an understanding of race as a technology in the sense of a specific technobiopolitical material-discursive apparatus of producing life and death through reading, categorizing, making intelligible, sorting, and hierarchizing bodies highlights that not only human knowledge and practices but also human and nonhuman bodies and agencies are actively involved in the materialization of racialized bodies. Understanding both the spirometer and UK Border Agency's *Human Provenance Pilot Project* as techno-apparatuses of bodily production illustrates this by demonstrating how bodies marked by race, ethnicity, and nationality emerge out of particular practices (which are by far not only

---

in allegedly more 'stable governments' in the so-called Western World compared to those in other parts of the world (Africa and the so-called Arabic World, in particular).

technological or technoscientific in nature), and in doing so that neither the category race nor racialized bodies are epistemologically or ontologically transcended ahistorical phenomena. Instead of being understood as preexisting, that is, as transcendental and supposedly natural phenomena or, on the contrary, as mere social and linguistic constructions, racialized bodies are reworked as effects of material-discursive relations; not matters of fact but matters of concern.

Such a move underlines the urgency of understanding questions of the body and its materialization always as questions of power by not only asking how specifically re(con)figured bodies come to matter but also by asking "which differences matter and for whom" (Barad 2007: 90). Against this backdrop, race could be further understood as a technology that aims "at permitting the exercise of biopower" (Mbembe 2003: 17) over bodies that have been marked. From such a point of view race appears to be closely linked to what Achille Mbembe termed necropolitics, that is, the "politics of death", functioning as a regulator for "the distribution of death" (ibid). And indeed, the history of the spirometer shows how bodies marked by race were not only regarded as less vital by an alleged 'natural law' but, even until today, also considered as not equally worth of compensation in the case of illness and disability compared to their white counterparts. Disability and death becomes technoscientifically calculated and embedded into the very instrument as well as into the data it produces. But even though the spirometer targets individual bodies, at the same time it is also directed at the technological and statistical control and administration of,

"populations that surface as profiles of bodily capacities, indicating what a body can do now and in the future. The affective capacity of bodies, statistically simulated as risk factors, can be apprehended as such without the subject, even without the individual subject's body, bringing forth competing bureaucratic procedures of control and political command in terms of securing the life of populations." (Clough 2008: 18)

This conjunction of estimating certain risk factors in the population with the political control over life and death represents a technique that, in the wake of Foucault, can be regarded as inherently racist because it aligns groups of the population along a hierarchized scale.

While the spirometer, as a techno-apparatus of bodily production, functions as a more traditional means for the exercise of power over populations, the case of the *Human Provenance Pilot Project* demonstrates that it would be a mistake to believe that with/in genomics race would become "less meaningful" on a subdermal or molecular scale; as, for example, the critical race scholar Paul Gilroy argued in the wake of the promises of the *Human Genome Project*.

"Genomics may send out the signal to reify 'race' as code and information, but there is a sense in which it also points unintentionally toward 'race's' overcoming. [...] At the smaller than microscope scales that open up the body for scrutiny today, 'race' becomes less meaningful, compelling, or salient to the basic tasks of healing and protecting ourselves." (Gilroy 2000: 37)

While genetics denotes the science of heredity which studies single genes in isolation, genomics refers to the study of the genome, that is, the entirety of the genetic material of an organism. After the use of genetics for eugenics, forced sterilization programs, and racial hygiene in the first half of the 20th century, genetics came under critique for its complicity in the crimes that had been committed against people stigmatized as religious and racialized others, LGBT people, Romani people and Irish Travellers, as well as disabled people – in short, against everyone who was constructed as so-called 'unworthy life'. In the wake of the discovery of the double-helix structure of the DNA in the early 1950s, a shift from the study of populations to individuals can be identified. After successfully sequencing the genomes of several bacteriophages (viruses that infect and replicate within bacteria) in the 1970s, and later also of bacteria and multicellular eukaryotes, in 2003 the *Human Genome Project* sequenced and mapped 99 percent of the human genome. As the totality of genes in organism, the genome is not a 'real thing', rather it has to be understood as "a thing-in-itself [...] not a whole in the traditional, 'natural' sense but a congeries of entities that are themselves autotelic and self-referential" (Haraway 1997: 134). The genome is "a historically new entity", something that is "radically 'culturally' produced" and yet "no less 'natural' for all that" (ibid: 148-149).

Jenny Reardon elaborates how genome scientist and policy makers such as the famous biochemist and geneticist Craig Venter have been arguing recently for the anti-racist potential of genomics, seeing genomics as "an antidote to racist ideology" (Reardon 2012: 26). Moving away from innate difference to a kind of bio-social assemblage which is characterized by plasticity instead of being fixed, this understanding of race is not the same as the one that dominated scientific, medical, and political debates at the turn to the 20th century. And yet, it seems that genomics "may create new forms of racism at the very moment that it explicitly seeks anti-racist ends", as Reardon (2012: 25) notes. Therefore, it seems that race today not only, once again, becomes a scientifically valid category in genomics but is also reinforced as a naturally occurring biological category. Taking into account the biopolitical history of the spirometer and the case of the *Human Provenance Pilot Project* demonstrates that genomics is far away from making race less meaningful. Instead, it could be said that today race does not only enter through the skin, as Frantz Fanon (1976) has put it with the notion of epidermalization, or Stuart Hall (2001) has seen it in the practices of the writing of difference on the skin of the other, but is increasingly read off of the very 'interior' of the body – that is, of the DNA, mitochondria, and the isotopes incorporated in the body, or the alleged 'efficiency' and function of certain organs. Samira Kawash puts this idea in a nutshell, stating,

"Race is on the skin, but skin is the sign of something deeper, something hidden in the invisible interior of the organism (as organic or ontological). To see racial difference is therefore to see the bodily sign of race but also to see more than this seeing, to see the interior difference it stands for." (Kawash 1997: 130)

Even though the UNESCO declaration on "The Race Question" from 1950, and its substantial revision from 1951, indicates that there is no evidence that would support the belief in race as a scientifically demarcated, naturally occurring category for grouping human beings, with the consequence that it would be wiser "to drop the term 'race' altogether" (ibid: 6),[50] scientific theories understanding race as a supposedly innate attribute of human bodies have not disappeared. On the contrary, it seems that the use of the category race has "increased during the past twenty years, after it had almost completely disappeared soon after the end of World War II" (Gissis 2008: 437). Ironically, after scientists of the *Human Genome Project* declared the death of race in biology and medicine in the year 2000 (cf., for example, Bliss 2012), biological understandings of race are currently flourishing again. This is especially visible in attempts toward a race-based medicine as well as the identification of DNA biomarkers that are believed to determine genetic disease. In a more than unsettling way, attempts toward a race-based medicine may run the risk of reviving what had been arduously shattered with the UNESCO statement on race after the Allied victory over the Axis powers, namely the "bioscientific tie of race, blood, and culture" (Haraway 1997: 239).

"If performativity is linked not only the formation of the subject but also to the production of the matter of bodies, as Butler's account of 'materialization'

---

50 While the first UNESCO statement on race declared that race "is not so much a biological phenomenon as a social myth" (Kuper 1975: 345), its revision from 1951 conceded that "it is possible, though not proved, that some types of innate capacity for intellectual and emotional response are commoner in one human group than another" (ibid: 352).

and Haraway's notion of 'materialized refiguration' suggest, then it is all the more important that we understand the nature of this production", Karen Barad (2003: 808) emphasizes. Exploring the biopolitical past and present of the spirometer and the case of the *Human Provenance Pilot Project* through the lens of the notion of the techno-apparatuses of bodily production provides us with an understanding of how difference materializes, that is, how differently reconfigured bodies come to matter, rather than only how different meanings are produced. From such a perspective, what constitutes race or, more precisely, racialized bodies, because race cannot do without the material body, are always shifting intra-active material-discursive forces in their multiple entanglements. In this light, radicalized bodies represent material-discursive phenomena, neither solid nor completely fluid, and never fully separable from the very techno-apparatuses of bodily production through which they come into being. "Materialization is never quite complete", as Judith Butler (1993: 2) urges us not to forget. Consequently, the bodies concerned, after their entanglement with the other forces and entities that constitute the spirometer and the *Human Provenance Pilot Project* as techno-apparatuses of bodily production, were not the same as they were before; they became bodies materially re(con)figured by race, ethnicity, sex, biology, politics, bodily efficiency, and geography. Both phenomena enact what matters and what is excluded from mattering in their own terms. What is more, both phenomena demonstrate what was meant in the previous chapter by emphasizing that not only bodies and identities but also their corresponding ontologies are not static, fixed, and essentially given but always fluid, always in becoming. "Boundaries do not sit still", as Karen Barad (2007: 171) puts it By reframing ontology as fluid, relational and immanent instead of as given, static, and transcendent ontology becomes a contested ground but also inherently political.

In shifting the focus onto processes of reconfigurings, rather than already given entities with fixed borders and inherent attributes which then somehow interact, connect, or influence one another – but nevertheless remain self-contained, stable entities with given essences – a new materialist account of the techno-apparatus of bodily production breaks with the repetition of the long-told story that divides the world into the spheres of the natural and the cultural, the organic and the mechanical, as well as matter and mind. With these dichotomies also their purpose, namely to function as a repository for the hierarchical construction of the world and the legitimation of oppression against "all constituted as others, whose task is to mirror the self" (Haraway 1991a: 177), could cease. It could even be said that with such an approach there is no longer either an essence of technology or an original, supposedly natural body *before* technology, as both are constantly reconfiguring each other – however, with very different political and ethical consequences, which precisely calls for the need of differentiating the productive attachments or entanglements from the potentially harmful ones.

It is for this reason that considering bodies as agentic and as not separable from the very techno-apparatuses of bodily production through which they come to matter and form part of, means to shift the focus from narratives emphasizing the power of technologies over material bodies and even over 'life itself' to the fact that also the most powerful and anti-democratic technologies can always fail; whereby to fail here precisely does not mean that the technologies and technoscientific practices involved were flawed or not yet powerful, invasive, or successful enough. In fact, such a perspective would be itself an instrumental and fatalistic one, leaving hardly any space for acknowledging political, biological, and material or thingly agency (as part of specific apparatuses). More importantly, such a perspective would remain caught in the logics of the myths of technological progress and hence in its own terms deterministic. As a result, such an account would only reproduce the very same belief in the omnipotence of science and technology that feminist and other critical scholars contested in their critique of positivist theories.

It is in this light that what a new materialist analysis of the biopolitical past and present of the spirometer and the case of the UK Border Agency's *Human Provenance Pilot Project* makes evident is that both phenomena are anything but smooth running technologies of control, functioning precisely as intended. As a matter of fact, both phenomena become instead visible as limited and fragile. What is contested with such a perspective is nothing less than the belief that certain technologies are so powerful and effective, that resistance would be literally futile (cf. Winner 2005).[51] A belief that often goes hand in hand with an understanding of society as pervaded thoroughly by control, dominance, and oppression. Even the most anti-democratic technologies also always generate resistance. In fact, they do not only 'generate' it, it always already forms a part of it. "Deterritorialized controls are far from perfect, however; they produce deterritorialized forms of resistance as a function of their own organization", as William Bogard (2006: 97) argues.[52] It is important to understand that resistance

---

51 See also page 110 and following pages.
52 While I tend to agree with William Bogard that technologies of control also produce or always already contain lines of flight in the sense of forces and flows that run counter to attempts of re-/territorialization, that is, acts of fixation and essentialization, I do not quite share his overly optimistic assumption that new technologies and, in particular, "the surveillance assemblage has opened a new deterritorialized space of commu-

(or, in a perhaps broader sense, agency), has not to be confused with *political* individual agency here but also comprises of *material* or *bodily* agency (which itself in turn is always the effect of multiplicities rather than a given property of specific entities). As pointed out earlier, in both cases theorized, agency marks the effect of entanglements of intra-acting human and nonhuman entities and forces: meaning, asylum seekers, case owner, politicians, geneticists, non-government organizations, institutions such as the UK Border Agency, technical instruments, political and economic discourses, global migration movements, ongoing wars and conflicts, isotopes, mitochondria, as well as the bodies under investigation themselves, to name but a few, in the case of the *Human Provenance Pilot Project*; physicians, plantation owners, black intellectuals, lawyers and courts, as well as technologies, statistics, statesmen, political discourses, nationalism, the political economy, racism, evolutionary theories, eugenics, and the organs and bodily functions under investigation, and much more, in the case of the spirometer. Consequently, both phenomena not only show how specifically reconfigured bodies material-discursively come to matter but also how the attempts at inscribing political interests, such as the intention to hinder undocumented migrants from entering the UK (or other countries for that matter), or attempts at hindering workers of color from applying for compensation payments for disability and illness into technologies did not function as intended. Once again, what becomes evident here is not only that it would be far too easy to understand technologies and technoscientific practices as condensed power relations or mere politics by other means but also that both human and nonhuman bodies are anything but passive objects. In fact, a new materialist account of bodies and technologies as outlined in this book suggests that in both cases material bodies proved to be resistant against the attempts of their technoscientific and biopolitical objectification, disrupting both the idea that bodies are mere passive objects that could be accessed and 'read-out' technologically, and the belief that there is a guarantee that technologies function precisely according to specific interests that have been embedded or inscribed into them.[53] As little as there is a guarantee that technologies and technoscientific practices function as

---

nication that with time may undermine the regime of global biopower" (Bogard 2006: 114).

53 Following Deleuze it could also be said that the forces and flows within these techno-apparatuses of bodily production could not be silenced and synchronized successfully, with the effect that the lines of flight succeeded over the territorializing attempts. However, Deleuze also reminds us that every line of flight brings also always with it new forms of territorialization.

intended, it would be epistemologically and politically limiting to understand the body first and foremost as a kind of "machine-readable information storage device" (van der Ploeg/Sprenkels 2011; Dijstelbloem et al. 2011a) or even a "truth machine" (Weiss 2011: 15). It is no wonder that bodies *seem* as technologically accessible objects if it is believed that there is no material body beyond its discursive and social construction. Taking in such a perspective negates any kind of material or bodily agency in advance and with it any possibilities to consider that even the most powerful technologies and technoscientific practices in their entanglement with material bodies can also fail, as I have stressed before. A new materialist account of the techno-apparatus of bodily production, on the contrary, opens up a perspective on material bodies and technologies in their multiple entanglements with one another beyond essentialism, determinism, and social constructionism; only this time material bodies and nonhuman entities have also something to say. To understand technologies as relational processes of reconfigurings, as a "relational matrix of humans and nonhumans" (Haraway 2007: 94) within which the human and the nonhuman, the organic and the mechanical, the biological and the technological, are deeply entangled with each other (yet, precisely, not in an ahistorical sense), suggests that there is no outside to technology. What is more, such an account contests the belief that material bodies, human and nonhuman ones alike, are only passive objects upon which certain technologies, policies, and practices act, inscribing socio-cultural norms, power relations, and politics into them (or, more precisely, onto their surfaces).

"There is no need to fear or hope, but only to look for new weapons", Gilles Deleuze (1992: 4) reminds us. Yet these weapons of thinking, these practices of developing new concepts and figures that, for example, allow us to take into account the potential of material bodies to be unruly and to "kick back" (Barad 2007: 215) are precisely what gets lost in theories that emphasize the omnipotence of science and technology. Acknowledging the potential of material bodies and things to 'kick back' does not mean to open up a dualism between technology and the body once again because both technologies and bodies are always multiplicities themselves which are in various ways entangled with one another. Moreover, acknowledging bodily agency, in an important sense, neither means to be indifferent against attempts of a technobiopolitical objectification, commodification, and reterritorialization of both human and nonhuman bodies, their forces, powers, and 'products', nor does it mean to fall into a romantic celebration of bodies and their potential to 'kick back' as agency can come in different forms and with very different consequences.

Shifting the focus toward the question what technologies do, rather than what technology is, what bodies can do, rather than what (the supposedly natural) body is, on the other hand, might provide us with an understanding of how technologies and material bodies in their entanglements with one another constitute the very material our world is made of; instead of only assuming that powerful invasive technologies would increasingly dominate, manipulate, and dissolve the body or even Nature itself (whatever that might be). The figure of the techno-apparatus of bodily production allows us to turn to these questions by taking in a different perspective that promises to see different things, or at least to see things differently. In doing so, it makes explicit what it means to cut technology and the body together-apart: namely, to disrupt beliefs in the omnipotence of technologies and technoscientific practices in order to be able "to work toward 'more promising interference patterns', both between words and things (allowing for things and bodies to be active in processes of signification)", as Iris van der Tuin (2011: 26-27) puts it. Even though, or perhaps, precisely because, agency is re-worked as an effect of entangled multiplicities, rather than being a property of the autonomous subject, techno-apparatuses of bodily production, as nexuses where power-knowledge-materiality concentrates and intra-actively materializes particular phenomena, are not determining; there is always the possibility for different entanglements with different worldly consequences. In doing so, what is opened up here is nothing less than an avenue that demonstrates that the potential to form different entanglements is always given but never determined and here also lies the possibility for more livable worlds.

References

Adorno, Theodor W. (2001 [1963]): Problems of Moral Philosophy, Stanford: Stanford University Press.
Adorno, Theodor W. (2003 [1953]): "Über Technik und Humanismus." In: Gesammelte Schriften Bd. 20.2, Frankfurt am Main: Campus. pp. 310-317.
Adorno, Theodor W. (2005 [1974]): Minima Moralia: Reflections on a Damaged Life, London and New York: Verso.
Adorno, Theodor W./Horkheimer, Max (1972 [1947]): The Dialectic of Enlightenment, New York: Herder and Herder.
Agamben, Giorgio (2000): Means Without End: Notes on Politics, Minneapolis: University of Minnesota Press.
Agamben, Giorgio (2005): What is a Dispositif? Lecture held at the European Graduate School. Transcript available online: (http://www.egs.edu/faculty/giorgio-agamben/articles/what-is-a-dispositif/part-1/) accessed March 15, 2013.
Ahmed, Sara (2008): "Imaginary Prohibitions: Some Preliminary Remarks on the Founding Gestures of the 'New Materialism'." European Journal of Women's Studies 15/1, pp. 23-39.
Akrich, Madeline (1992): "The De-Scription of Technical Objects." In: Wiebe E. Bijker/John Law (eds.), Shaping Technology/Building Society: Studies in Sociotechnical Change, Cambridge, Mass.: MIT Press, pp. 205-224.
Anders, Günther (1956): Die Antiquiertheit des Menschen: Über die Seele im Zeitalter der zweiten industriellen Revolution, München: C.H. Beck.
Anders, Günther (1980): Die Antiquiertheit des Menschen. Bd. 2: Über die Zerstörung des Lebens im Zeitalter der dritten industriellen Revolution, München: C.H. Beck.
Arendt, Hannah (2000): The Portable Hannah Arendt, ed. by Peter Baehr, New York: Penguin Books.

Åsberg, Cecilia (2008): "A Feminist Companion to Post-humanities." NORA— Nordic Journal of Feminist and Gender Research 16/4, pp. 246-269.

Åsberg, Cecilia/Birke, Lynda (2010): "Biology is a Feminist Issue: Interview with Lynda Birke." European Journal of Women's Studies 17/4, pp. 413-423.

Bagchi, Arjun/Detournay, Stephane/Grumiller, Daniel/Simón, Joan (2013): "Cosmic Evolution from Phase Transition of Three-Dimensional Flat Space." Physical Review Letters 111(18):181301, pp. 1-5.

Balsamo, Anne (1995): "Forms of Technological Embodiment." In: Mike Featherstone/Roger Burrows (eds.), Cyberspace/Cyberbodies/Cyberpunk: Cultures of Technological Embodiment, London and Thousand Oaks: Sage, pp. 215-237.

Balsamo, Anne (1996): Technologies of the Gendered Body: Reading Cyborg Women, Durham and London: Duke University Press.

Barad, Karen (2001): "Re(con)figuring Space, Time, and Matter." In: Marianne DeKoven (ed.), Feminist Locations: Global and Local: Theory and Practice, New Brunswick: Rutgers University Press, pp. 75-109.

Barad, Karen (2003): "Posthumanist Performativity: Toward an Understanding of How Matter Comes to Matter." Signs: Journal of Women in Cultural and Society 28/3, pp. 801-831.

Barad, Karen (2007): Meeting the Universe Halfway: Quantum Physics and the Entanglement of Matter and Meaning, Durham and London: Duke University Press.

Barad, Karen (2011): "Erasers and Erasures: Pinch's Unfortunate 'Uncertainty Principle'." Social Studies of Science 41/3, pp. 443-454.

Barad, Karen (2012a): "Intra-Active Entanglements: An Interview with Karen Barad." KVINDER, KØN & FORSKNING (Special Issue on Feminist Materialisms) 21/1-2, pp. 10-21.

Barad, Karen (2012b): What is the Measurement of Nothingness? Infinity, Virtuality, Justice, dOCUMENTA (13), Ostfildern: Hatje Cantz.

Barad, Karen (2012c): "'Matter Feels, Converses, Suffers, Desires, Yearns and Remembers': Interview with Karen Barad." In: Rick Dolphijn/Iris van der Tuin (eds.), New Materialism: Interviews & Cartographies, Ann Arbor, MI: Open Humanities Press, pp. 48-70.

Barad, Karen (2014): "Diffracting Diffractions: Cutting Together-Apart." Parallax (Special Issue: Diffracted Worlds – Diffractive Readings: Onto-Epistemologies and the Critical Humanities) 20/3, pp. 168-187.

Barla, Josef (2016). "Technologies of Failure, Bodies of Resistance: Science, Technology and the Mechanics of Materializing Marked Bodies." In Victoria

Pitts-Taylor (ed.), Mattering. Feminism, Science, and Materialism, New York: New York University Press, pp. 159-172.

Barla, Josef/Steinschaden, Fabian (2012): "Kapitalistische Quasi-Objekte. Zu einer Latour'schen Lesart Marx' Ausführungen zur Maschine." In: Alfred Dunshirn/Elisabeth Nemeht/Gerhard Unterthurner (eds.), Crossing Borders: Thinking (across) Boundaries, Wien: œgp, pp. 361-370.

Barla, Josef/Hubatschke, Christoph (2017). "Technoecologies of Borders: Thinking with Borders as Multispecies Matters of Care." Australian Feminist Studies 32/94: 395-410.

Barnes, Barry (1974): Scientific Knowledge and Social Theory, London and Boston: Routledge & Kegan Paul.

Barnes, Barry/Bloor, David (1982): "Relativism, Rationalism and the Sociology of Knowledge." In: Martin Hollis/Steven Lukes (eds.), Rationality and Relativism, Cambridge, MA: MIT Press, pp. 21-47.

Barthélémy, Jean-Hughes (2011): "Simondon – Ein Denken der Technik im Dialog mit der Kybernetik." In: Erich Hörl (ed.), Die technologische Bedingung. Beiträge zur Beschreibung der technischen Welt, Frankfurt am Main: Suhrkamp, pp. 93-109.

Becker, Egon/Jahn, Thomas (2003). "Umrisse einer kritischen Theorie gesellschaftlicher Naturverhältnisse." In: Gernot Böhme/Alexandra Manzei (eds.), Kritische Theorie der Technik und der Natur, München: Fink, pp. 91-112.

Becker-Schmidt, Regina (2000): Feministische Theorien zur Einführung, Hamburg: Junius.

Benjamin, Walter (2008 [1936]): The Work of Art in the Age of Mechanical Reproduction. New York: Penguin Books.

Benjamin, Walter (2014): Radio Benjamin, ed. by Lecia Rosenthal, London and New York: Verso.

Bentham, Jeremy (1995): The Panopticon Writings, ed. by Miran Božovič, London: Verso.

Berg, Anne-Jorunn/Lie, Merete (1995): "Feminism and Constructivism: Do Artifacts Have Gender?" Science, Technology, & Human Values 20/3, pp. 332-351.

Berger, Peter/Luckmann, Thomas (1967): The Social Construction of Reality: A Treatise in the Sociology of Knowledge, London: Penguin Books.

Bérubé, Michael (2011): "The Science Wars Redux." Democracy: A Journal of Ideas 19 (Winter), pp. 64-74.

Bijker, Wiebe E./Hughes, Thomas/Pinch, Trevor (eds.) (1987): The Social Construction of Technological Systems: New Directions in the Sociology and History of Technology, Cambridge, MA: MIT Press.

Bijker, Wiebe E./Law, John (eds.) (1992): Shaping Technology/Building Society: Studies in Sociotechnical Change, Cambridge, MA: MIT Press.
Bijker, Wiebe E. (1992): "The Social Construction of Fluorescent Lightning or How an Artifact Was Invented in Its Diffusion Stage." In: Wiebe E. Bijker/John Law (eds.), Shaping Technology/Building Society: Studies in Sociotechnical Change, Cambridge, MA: MIT Press, pp. 75-102.
Bijker, Wiebe E. (2006): "Why and How Technology Matters." In: Robert Goodin and Charles Tilly (eds.), The Oxford Handbooks of Political Science, Oxford and New York: Oxford University Press, pp. 681-706.
Bimber, Bruce (1990): "Karl Marx and the Three Faces of Technological Determinism." Social Studies of Science 20/2, p. 333-351.
Birke, Lynda (1999): Feminism and the Biological Body, Edinburgh: Edinburgh University Press.
Birke, Lynda/Hubbard, Ruth (eds.) (1995): Reinventing Biology: Respect for Life and the Creation of Knowledge, Bloomington: Indiana University Press.
Bishop, P.J. (1977): "A Bibliography of John Hutchinson." Medical History 21/4, pp. 384-396.
Bliss, Catherine (2012): Race Decoded: The Genomic Fight for Social Justice, Stanford: Stanford University Press.
Bloor, David (1976): Knowledge and Social Imagery, London and Boston: Routledge & Kegan Paul.
Bloor, David (1991): Knowledge and Social Imagery, Second Edition, London: University of Chicago Press.
Bogard, William (2006): "Surveillance Assemblages and Lines of Flight." In: David Lyon (ed.), Theorizing Surveillance: The Panopticon and Beyond, Cullompton and Portland: Willan, pp. 97-122.
Bohr, Niels (2005): Collected Works, Vol. 11: The Political Arena (1943-1961), ed. by Finn Aaserud, Amsterdam: Elsevier.
Böhme, Gernot (2012): Invasive Technification: Critical Essays in the Philosophy in Technology, London: Bloomsbury.
Bostrom, Nick (2003): The Transhumanist FAQ: A General Introduction, (http://www.transhumanism.org/resources/FAQv21.pdf) accessed November 12, 2012.
Braidotti, Rosi (2011): Nomadic Subjects: Embodiment and Sexual Difference in Contemporary Feminist Theory, New York: Columbia University Press.
Braidotti, Rosi (2013): The Posthuman, Cambridge and Malden: Polity Press.
Braun, Kathrin (1998): "Mensch, Tier, Chimäre: Grenzauflösung durch Technologie." In: Gudrun-Axeli Knapp (ed.), Kurskorrekturen: Feminismus

zwischen Kritischer Theorie und Postmoderne, Frankfurt am Main and New York: Campus, pp. 153-177.

Braun, Lundy (2005): "Spirometry, Measurement, and Race in the Nineteenth Century." Journal of the History of Medicine and Allied Science 60/2, pp. 135-169.

Braun, Lundy (2014a): Breathing Race into the Machine: The Surprising Career of the Spirometer from Plantation to Genetics, Minneapolis and London: University of Minnesota Press.

Braun, Lundy (2014b) "How Racism Creeps Into Medicine." In: The Atlantic August 29, (http://www.theatlantic.com/health/archive/2014/08/how-racism-creeps-into-medicine/378618/) accessed September 10, 2017

Browne, Simone (2012): "Race and Surveillance" In: Kristie Ball/Kevin Haggerty/David Lyon (eds.), Routledge Handbook of Surveillance Studies, London and New York: Routledge, pp. 72-79.

Bruining, Dennis (2013): "A Somatechnics of Moralism: New Materialism or Material Foundationalism" Somatechnics 3/1, pp. 140-168.

Butler, Judith (1993): Bodies That Matter: On the Discursive Limits of 'Sex', New York and London: Routledge.

Butler, Judith (2004): Undoing Gender, New York and London: Routledge.

Butler, Judith (2012): "Can One Lead a Good Life in a Bad Life? Adorno Price Lecture" Radical Philosophy 178, pp. 9-18.

Callon, Michel (1986): "The Sociology of an Actor-Network: The Case of the Electric Vehicle" In: Michel Callon/John Law/Arie Rip (eds.), Mapping the Dynamics of Science and Technology: Sociology of Science in the Real World, Basingstoke: Macmillan, pp. 19-34.

Cassidy, David C. (2009): Beyond Uncertainty: Heisenberg, Quantum Physics, and the Bomb, New York: Bellevue Literary Press.

Chabot, Pascal (2013): The Philosophy of Simondon: Between Technology and Individuation, London and New York: Bloomsbury.

Chorley, Matt/Slack, James/Chapman, James (2013): "'Immigration System is Like a Never-Ending Game of Snakes and Ladders': Theresa May Vows to Kick Out Illegal Migrants Before They Get Chance to Appeal". In: This is Money September 29, (http://www.thisismoney.co.uk/news/article-2438130/Theresa-May-Ill-kick-illegal-migrants-BEFORE-chance-appeal.html/) accessed October 27, 2013.

Chun, Wendy Hui Kyong (2009): "Introduction: Race and/as Technology; or, How to Do Things to Race" Camera Obscura 24/1, pp. 6-35.

Clarke, Adele E. (1995): "Modernity, Postmodernity & Reproductive Processes ca. 1890-1990. Or, 'Mommy, where do cyborgs come from anyway?'." In:

Chris Hables Gray (ed.), The Cyborg Handbook, New York and London: Routledge, pp. 139-155.

Clough, Patricia (2008): "The Affective Turn: Political Economy, Biomedia and Bodies" Theory, Culture & Society 25/1, pp. 1-22.

Clynes, Manfred E./Kline, Nathan S. (1960): "Cyborgs and Space." Astronautics September 1960, pp. 26-27, 74-76.

Clynes, Manfred E./Gray, Chris Hables (1995): "An Interview with Manfred Clynes." In: Chris Hables Gray (eds.), The Cyborg Handbook, New York and London: Routledge, pp. 43-53.

Cockburn, Cynthia (1985): "The Material of Male Power." In: Donal MacKenzie/Judy Wajcman (eds.), The Social Shaping of Technology, Milton Keynes: Open University Press, pp. 125-146.

Cockburn, Cynthia/Dilic, Ruza (eds.) (1994): Bringing Technology Home: Gender and Technology in Changing Europe, Buckingham: Open University Press.

Coleman, Beth (2009): "Race as Technology" Camera Obscura 70/24:1, pp. 176-207.

Collins, Harry (1995): "Humans, Machines, and the Structure of Knowledge.", Stanford Electronic Humanities Review 4/2, (http://www.stanford.edu/group/SHR/) accessed August 9, 2018.

Collins, Harry/Kusch, Martin (1998): The Shape of Actions: What Humans and Machines Can Do, Cambridge, MA: MIT Press.

Combes, Muriel (2013): Gilbert Simondon and the Philosophy of the Transindividual, Cambridge, MA and London: MIT Press.

Davis, Kathy (2007): "Reclaiming Women's Bodies: Colonialist Trope or Critical Epistemology?" The Sociological Review 55 (Supplement s1), pp. 50-64.

Davis, Noela (2009): "New Materialism and Feminism's Anti-Biologism: A Response to Sara Ahmed." European Journal of Women's Studies 16/1, pp. 67-80.

Davis, Noela (2014): "Politics Materialized: Rethinking the Materiality of Feminist Political Action through Epigenetics." Women: A Cultural Review 25/1, pp. 62-77.

de Boever, Arne/Murray, Alex/Roffe, Jon/Woodward, Ashley (eds.) (2013): Gilbert Simondon on Being and Technology, Edinburgh: Edinburgh University Press.

de Broglie, Louis (1939): Matter and Light: The New Physics. New York: W.W. Norton.

Deleuze, Gilles (1983): Nietzsche and Philosophy, New York: Columbia University Press.

Deleuze, Gilles (1992): "Postscript on the Societies of Control." OCTOBER 59, Cambridge, MA: MIT Press, pp. 3-7.

Deleuze, Gilles (1995): Negotiations, New York: Columbia University Press.

Deleuze, Gilles (2003): Francis Bacon: The Logic of Sensation, London and New York: Continuum.

Deleuze, Gilles/Guattari, Felix (2004): A Thousand Plateaus. New York and London: Continuum.

Deleuze, Gilles/Parnet, Claire (1977): Dialogues, New York: Columbia University Press.

Derrida, Jacques (1995): *Points.... Interviews, 1974–1994*, ed. by Elisabeth Weber, Stanford: Stanford University Press.

Derrida, Jacques (2012): Specters of Marx, New York and London: Routledge.

Dijstelbloem, Huub/Meijer, Albert/Besters, Michiel (2011a): "The Migration Machine." In: Huub Dijstelbloem/Albert Meijer (eds.), Migration and the New Technological Borders of Europe, Basingstoke: Palgrave Macmillan, pp. 1-21.

Dijstelbloem, Huub/Meijer, Albert/Broms, Frans (2011b): "Reclaiming Control over Europe's Technological Borders." In: Huub Dijstelbloem/Albert Meijer (eds.), Migration and the New Technological Borders of Europe, Basingstoke: Palgrave Macmillan, pp. 170-185.

Dolphijn, Rick/van der Tuin, Iris (2012): New Materialism: Interviews & Cartographies, Ann Arbor, MI: Open Humanities Press.

Douglas, Frederick (2000 [1854]): "The Claims of the Negro Ethnologically Considered." In: Frederick Douglass, Selected Speeches and Writings, ed. by Philip S. Foner/Yuval Taylor, Chicago: Lawrence Hill Books, pp. 282-297.

Du Bois, W. E. B. (1898): "The Study of the Negro Problems." The Annals of the American Academy of Political and Social Science 11 (January), pp. 1-23.

Duden, Barbara (1993): Disembodying Women. Perspectives on Pregnancy and the Unborn, Cambridge, MA: Harvard University Press.

Dusek, Val (2006): Philosophy of Technology: An Introduction, Malden and Oxford: Blackwell.

Duster, Troy (2006). "Explaining Differential Trust of DNA Forensic Technology: Grounded Assessment or Inexplicable Paranoia?" The Journal of Law, Medicine & Ethics 34/2, pp. 293-300.

Eagleton, Terry (2003): After Theory, London and New York: Basic Books.

Engels, Friedrich (1978 [1872]): "On Authority" In: The Marx Engels Reader, ed. by Robert C. Tucker, New York and London: Norton, pp. 730-733.

Engels, Friedrich (2005 [1845/1892]): The Condition of the Working-Class in England in 1844, (http://www.gutenberg.org/files/17306/17306-h/17306-h.htm) accessed December 20, 2012.

Fanon, Frantz (1967): Black Skin, White Mask, New York: Grove Press.

Farraj, Achraf (2011): "Refugees and the Biometric Future: The Impact of Biometrics on Refugees and Asylum Seekers." Columbia Human Rights Law Review 42/3, pp. 891-944.

Faulkner, Wendy (2001): "The Technology Question in Feminism: A View From Feminist Technology Studies." Women's Studies International Forum 24/1, pp. 79-95.

Faulkner, Wendy/Lohan, Maria (2004): "Masculinities and Technology. Some Introductory Remarks." Men and Masculinities 6/4, pp. 319-329.

Fausto-Sterling, Anne (2008): "The Bare Bones of Race." Social Studies of Science 38/5, pp. 657-694.

Feenberg, Andrew (2003): "Democratic Rationalization: Technology, Power, and Freedom." In: Robert C. Scharff/Val Dusek (eds.), Philosophy of Technology: The Technological Condition, Malden and Oxford: Blackwell, pp. 653-655.

Feenberg, Andrew (2004): *Questioning Technology: Interview with Roy Christopher*. October 12, (http://roychristopher.com/andrew-feenberg-questioning-technology/) accessed November 23, 2013.

Feenberg, Andrew (2010): "The Critical Theory of Technology." In: Craig Hanks (ed.), Technology and Values. Essential Readings, Malden and Oxford: Blackwell.

Fisher, Neil (2008): "Biometrics – It's Not What You Know, It's Who You Are!" Biometric Technologies Today 6/11-12, pp. 7-9.

Fleck, Ludwik (1935): Entstehung und Entwicklung einer wissenschaftlichen Tatsache: Einführung in die Lehre vom Denkstil und Denkkollektiv, Basel: Schwabe.

Fleck, Ludwik (1979): Genesis and Development of a Scientific Fact, ed. by Thaddeus J. Trenn/Robert K. Merton, Chicago and London: The University of Chicago Press.

Foucault, Michel (1972): The Archeology of Knowledge, London: Tavistock Publications.

Foucault, Michel (1977): Discipline and Punish: The Birth of the Prison, New York: Vintage Books.

Foucault, Michel (1978): The History of Sexuality. Volume 1: An Introduction, New York: Pantheon Books.

Foucault, Michel (1980): Power/Knowledge: Selected Interviews & Other Writings 1972-1977, New York: Pantheon Books.
Foucault, Michel (1984): "Space, Knowledge, and Power." In: Paul Rabinow (ed.), The Foucault Reader, New York: Pantheon Books, pp 239-256.
Foucault, Michel (1988): Technologies of the Self: A Seminar with Michel Foucault, ed. by Luther H. Martin, Huck Gutman, and Patrick H. Hutton, Amherst, MA: University of Massachusetts Press.
Foucault, Michel (2003): "Society Must Be Defended." Lectures at the Collège de France, 1975-76, ed. by François Ewald/Alessandro Fontana. New York: Picador.
Gane, Nicholas/Haraway, Donna (2006): "When We Have Never Been Human, What Is to Be Done?: Interview with Donna Haraway." Theory, Culture & Society 23/7-8, pp. 135-158.
Gibson, G.J. (2005): "Spirometry: then and now." Breath 1/3, pp 206-216.
Gilroy, Paul (2000): Against Race: Imagining Political Culture Beyond the Color Line, Cambridge, MA: Harvard University Press.
Gissis, Snait B. (2008): "When is 'Race' a Race? 1946–2003." Studies in History and Philosophy of Biology and Biomedical Sciences 39/4, pp. 437-450.
Gray, Chris Hables/Mentor, Steven/Figueroa-Sarriera, Heidi J. (1995): "Introduction: Constructing the Knowledge of Cybernetic Organisms." In: Chris Hables Gray (ed.), The Cyborg Handbook, New York and London: Routledge, pp. 1-14.
Greely, Henry/Riordan, Daniel/Garrison, Nanibaa/Mountain, Joanna (2006): "Family Ties: The Use of DNA Offender Databases to Catch Offenders' Kin." Journal of Law, Medicine & Ethics 34/2, pp. 248-262.
Griffin, Carl (2012): The Rural War, Manchester: Manchester University Press.
Grosz, Elizabeth (2004): The Nick of Time: Politics, Evolution, and the Untimely, Durham and London: Duke University Press.
Guthman, Julie/Mansfield, Becky (2012): "The Implications of Environmental Epigenetics: A New Direction of Geographic Inquiry on Health, Space, and Nature-Society Relation." Progress in Human Geography 37/4, pp. 486-504.
Habermas, Jürgen (1970): "Technology and Science as 'Ideology'." In: Toward a Rational Society, Boston: Beacon Press, pp. 91-122.
Habermas, Jürgen (1983): Philosophical-Political Profiles, Cambridge, MA: MIT Press.
Habermas, Jürgen (2003): The Future of Human Nature, Cambridge and Oxford: Polity Press.

Hacking, Ian (1982): "Biopower and the Avalanche of Printed Numbers." Humanities in Society 5/3-4, pp. 279-295.
Hacking, Ian (1998): "Canguilhem Amid the Cyborgs" Economy and Society 27/2-3, pp. 202-216.
Hacking, Ian (1999): The Social Construction of What?, Cambridge, MA: Harvard University Press.
Hacking, Ian (2009): "The Abolition of Man." Behemoth. A Journal of Civilisation 2/3, pp. 5-23.
Hall, Stuart (2001): The Multicultural Question, Milton Keynes: Pavis Centre for Social and Cultural Research at The Open University.
Haraway, Donna (1988): "Situated Knowledges: The Science Question in Feminism and the Privilege of Partial Perspective." Feminist Studies 14/3, pp. 575-599.
Haraway, Donna (1989): Primate Visions: Gender, Race and Nature in the World of Modern Science, London and New York: Routledge.
Haraway, Donna (1991a): Simians, Cyborgs, and Women: The Reinvention of Nature, New York: Routledge.
Haraway, Donna (1991b): "Cyborgs at Large: Interview with Donna Haraway." In: Constance Penley/Andrew Ross (eds.), Technoculture, Minneapolis: University of Minnesota Press, pp. 1-20.
Haraway, Donna (1992): "The Promises of Monsters: A Regenerative Politics for Inappropriate/d Others." In: Lawrence Grossberg/Cary Nelson/Paula Treichler (eds.), Cultural Studies, New York and London: Routledge, pp. 295-337.
Haraway, Donna (1994): "A Game of Cat's Cradle: Science Studies, Feminist Theory, Cultural Studies." Configurations: A Journal of Literature and Science 2/1, pp. 59-71.
Haraway, Donna (1995a): "Wir sind immer mittendrinnen: Ein Interview mit Donna Haraway." In: Die Neuerfindung der Natur: Primaten, Cyborgs und Frauen, Frankfurt am Main and New York: Campus, pp. 98-122.
Haraway, Donna (1995b) "Cyborgs and Symbionts: Living Together in the New World Order." In: Chris Hables Gray (ed.), The Cyborg Handbook, New York and London: Routledge, xi-xx.
Haraway, Donna (1997): Modest_Witness@Second_Millenium.FemaleMan© _Meets_OncoMouse™: Feminism and Technoscience, London and New York: Routledge.
Haraway, Donna (2000): How Like a Leaf: An Interview with Thyrza Nichols Goodeve, New York: Routledge.

Haraway, Donna (2003a): "Interview with Donna Haraway." In: Don Ihde/Evan Selinger (eds.), Chasing Technoscience: Matrix for Materiality, Bloomington and Indianapolis: Indiana University Press, pp. 47-57.
Haraway, Donna (2003b): The Companion Species Manifesto. Dogs, People, and Significant Otherness, Chicago: Prickly Paradigm Press.
Haraway, Donna (2004): The Haraway Reader, New York and London: Routledge.
Haraway, Donna (2005): "Conversations with Donna Haraway." In: Joseph Schneider, *donna haraway: live theory*, London and New York: Continuum, pp. 114-156.
Haraway, Donna (2007): "Five Questions." In: Jan-Kyrre Berg Olsen/Evan Selinger (eds.), Philosophy of Technology: Five Questions, New York: Automatic Press/VIP, pp. 91-99.
Haraway, Donna (2008): When Species Meet, Minneapolis and London: University of Minnesota Press.
Haraway, Donna (2011a): *SF: Science Fiction, Speculative Fabulation, String Figures, So Far, Pilgrim Award Acceptance Comments*, (http://people.ucsc. edu/haraway/Files/PilgrimAcceptance-Haraway.pdf) accessed January 30, 2014.
Haraway, Donna (2011b): *SF: Speculative Fabulation and String Figures/SF: Spekulative Fabulation und String-Figuren*, Ostfildern: Hatje Cantz.
Haraway, Donna (2016): Manifestly Haraway, Minneapolis and London: University Press of Minnesota Press.
Haraway, Donna/Williams, Jeffrey (2009): "Donna Haraway's Critters." In: The Chronicle of Higher Education October 18. (http://chronicle.com/article/A-Theory-of-Critters-/48802/) accessed April 8, 2013.
Harbers, Hans (2003): "The Womb as Operation Room: Feminist Technology Studies Without 'Failures of Nerve'." Science, Technology & Human Values 28/3, pp. 425-434.
Harding, Sandra (2008): Sciences From Below: Feminisms, Postcolonialities, and Modernities, Durham and London: Duke University Press.
Harrasser, Karin (2006): "Donna Haraway: Natur-Kulturen und die Faktizität der Figuration." In: Stephan Moebius/Dirk Quadflieg (eds.), Kultur. Theorien der Gegenwart, Wiesbaden: VS Verlag für Sozialwissenschaften, pp. 580-594.
Heidegger, Martin (1977): The Question Concerning Technology and Other Essays, New York and London: Garland Publishing.
Heidegger, Martin (1979): Nietzsche, Vol 1: The Will to Power as Art, New York: Harper and Row.

Heidegger, Martin (1992 [1942/43]): Parmenides, Bloomington and Indianapolis: Indiana University Press.
Heilbroner, Robert (1967): "Do Machines Make History?" Technology and Culture 8/3, pp. 335-345.
Heisenberg, Werner (1958): Physics and Philosophy: The Revolution in Modern Science, New York: Harper.
Hessen, Boris (2009 [1931]): "The Social and Economic Roots of Newton's Principia." In: Gideon Freudenthal/Peter McLaughlin (eds.), The Social and Economic Roots of the Scientific Revolution: Texts by Boris Hessen and Henryk Grossmann, New York: Springer, pp. 41-101.
Hopkins, Gail (2006): "Somali Community Organizations in London and Toronto: Collaboration and Effectiveness." Journal of Refugee Studies 19/3, pp. 361-380.
Hughes, Thomas (1986): "The Seamless Web: Technology, Science, Etcetera, Etcetera." Social Studies of Science 16, pp. 281-292.
Hughes, Thomas (1987): "The Evolution of Large Technological Systems." In: Wiebe E. Bijker/Thomas Hughes/Trevor Pinch (eds.), The Social Construction of Technological Systems: New Directions in the Sociology and History of Technology, Cambridge. MA: MIT Press, pp. 51-82.
Hughes, Thomas (2005): Human-Built World: How to Think About Technology and Culture, Chicago and London: University of Chicago Press.
Hutchinson, John (1846): "On the Capacity of the Lungs, and on the Respiratory Functions, With a View of Establishing a Precise and Easy Method of Detecting Disease by the Spirometer." Medico-Chirurgical Transactions 29, pp. 137-252. (https://www.ncbi.nlm.nih.gov/pmc/articles/PMC2116876/pdf/medcht00053-0183.pdf) accessed September 2, 2017.
Ihde, Don (2010): Heidegger's Technologies: Postphenomenological Perspectives, New York: Fordham University Press.
Jackson, Lee (2015): Dirty Old London: The Victorian Fight Against Filth, New Haven: Yale University Press.
Joerges, Bernward (1999): "Do Politics Have Artefacts?" Social Studies of Science 29/3, p. 411-431.
Kant, Immanuel (1951 [1790]): Critique of Judgement, New York: Hafner Press.
Kawash, Samira (1997): Dislocating the Color Line: Identity, Hybridity, and Singularity in African-American Narrative, Stanford: Stanford University Press.
Khosravi, Shahram (2010): 'Illegal' Traveller: An Auto-Ethnography of Borders, Basingstoke: Palgrave.

King, Katie (1991): "Bibliography and a Feminist Apparatus of Literary Production." Text: Transactions of the Society for Textual Scholarship 5, pp. 91-103.

Kirby, Vicki (1997): Telling Flesh: The Substance of the Corporeal, New York and London: Routledge.

Kjellén, Rudolf (1916): Staten som Lifsform, Stockholm: Hugo Grebers Förlag.

Kunzru, Hari (1997): "You are Cyborg." In: Wired Magazine February 1, (https://www.wired.com/1997/02/ffharaway/) accessed November 14, 2018.

Kuper, Leo (1975) Race, Science and Society, Paris and New York: The UNESCO Press and Columbia University Press.

Latour, Bruno (1987): Science in Action: How to Follow Scientists and Engineers through Society, Cambridge, MA: Harvard University Press.

Latour, Bruno (1988): The Pasteurization of France, Cambridge, MA and London: Harvard University Press.

Latour, Bruno (1991a): "Technology is Society Made Durable." In: John Law (ed.), A Sociology of Monsters: Essays on Power, Technology and Domination, London: Routledge, pp. 103-132.

Latour, Bruno (1991b): "The Berlin Key or How to Do Words with Things." In: Paul Graves-Brown (ed.), Matter. Materiality and Modern Culture, London: Routledge, pp. 10-21.

Latour, Bruno (1993): We Have Never Been Modern, Cambridge, MA: Harvard University Press.

Latour, Bruno (1998): "To Modernize or to Ecologize? That's the Question." In: Bruce Braun/Noel Castree (eds.), Remaking Reality: Nature at the Millenium, London and New York: Routledge, pp. 221-242.

Latour, Bruno (1999a): Pandora's Hope: Essays in the Reality of Science Studies, Cambridge, MA: Harvard University Press.

Latour, Bruno (1999b): "Factures/Fractures. From the Concept of Network to the Concept of Attachment." RES: Anthropology and Aesthetics 36, pp. 20-31.

Latour, Bruno (2000): "Die Kühe haben das Wort", In: Die Zeit November 30, (http://www.zeit.de/2000/49/Die_Kuehe_haben_das_Wort) accessed September 27, 2018.

Latour, Bruno (2002): "Morality and Technology: The End of the Means." Theory, Culture and Society 19/5-6 pp. 247-260.

Latour, Bruno (2003): "Interview with Bruno Latour (with Robert Crease, Don Ihde, Casper Bruun Jensen, and Evan Selinger)." In: Don Ihde/Evan Selinger (eds.), Chasing Technoscience: Matrix for Materiality, Bloomington and Indianapolis: Indiana University Press, pp. 15-26.

Latour, Bruno (2004a): "How to Talk About the Body? The Normative Dimension of Science Studies." Body & Sociology 10/2-3, pp. 205-229.
Latour, Bruno (2004b): "Why Has Critique Run out of Steam? From Matters of Fact to Matters of Concern." Critical Inquiry 20/2, pp. 225-248.
Latour, Bruno (2004c): Politics of Nature: How to Bring the Sciences Into Democracy. Cambridge, MA: London: Harvard University Press.
Latour, Bruno (2005a): Reassembling the Social, Oxford: Oxford University Press.
Latour, Bruno (2005b): "From Realpolitik to Dingpolitik or How to Make Things Public." In: Bruno Latour/Peter Weibel (eds.), Making Things Public: Atmospheres of Democracy. Cambridge, MA: MIT Press, pp. 14-43.
Latour, Bruno (2007a): "Turning Around Politics: A Note on Gerard de Vries' Paper." Social Studies of Science 37/5, pp. 811-820.
Latour, Bruno (2007b): "Can We Have Our Materialism Back, Please?" Isis 98/1, pp. 138-142.
Latour, Bruno (2013): An Inquiry into Modes of Existence: An Anthropology of the Moderns, Cambridge, MA and London: Harvard University Press.
Latour, Bruno/Woolgar, Steve (1979): Laboratory Life: The Social Construction of Scientific Facts, Los Angeles: Sage.
Latour, Bruno/Crawford, Hugh T. (1993): "An Interview with Bruno Latour." Configurations 1/2, pp. 247-268.
Latour, Bruno/Hermant, Emilie (1998): Paris ville invisible, La Découverte: Paris. English translation available via (http://www.bruno-latour.fr/sites/ default/files/downloads/ viii_paris-city-gb.pdf) accessed March 15, 2013.
Latour, Bruno/Sánches-Criado, Tomás (2007): "Making the 'Res Public'." Ephemera: Theory & Politics in Organization 7/2, pp. 364-371.
Law, John (1992): "Notes on the Theory of the Actor-Network: Ordering, Strategy, and Heterogeneity." Systems Practices 5/4, pp. 379-393.
Law, John. 2008. "Actor Network Theory and Material Semiotics." In Bryan S. Turner (ed.): *The New Blackwell Companion to Social Theory*. Malden and Oxford: Wiley-Blackwell, 141-158.
Law, John (2012): "Collateral Realities." In: Fernando Dominguez Rubio/Patrick Baert (eds.), The Politics of Knowledge, London: Routledge, pp. 158-178.
Law, John/Bijker, Wiebe E. (1992): "Postscript: Technology, Stability, and Social Theory." In: Wiebe E. Bijker/John Law (eds.), Shaping Technology/Building Society: Studies in Sociotechnical Change, Cambridge, MA: MIT Press, 290-308.

Law, John/Callon, Michel (1992): "The Life and Death of an Aircraft: A Network Analysis of Technical Change." In: Wiebe E. Bijker/John Law (eds.), Shaping Technology/Building Society. Studies in Sociotechnical Change, Cambridge, MA: MIT Press, pp. 21-52.

Leapman, Ben (2006): "Three in Four Young Black Men on the DNA Database." In: *The Telegraph* November 5, (http://www.telegraph.co.uk/news/uk news/1533295/Three-in-four-young-black-men-on-the-DNA-database.html/) accessed April 17, 2012.

Lemke, Thomas (2011): Biopolitics: An Advanced Introduction, New York and London: New York University Press.

Lemke, Thomas (2013): Die Natur in der Soziologie: Gesellschaftliche Voraussetzungen und Folgen biotechnologischen Wissens, Frankfurt am Main: Campus.

Lemke, Thomas (2015): "New Materialisms: Foucault and the 'Government of Things'." Theory, Culture & Society 32/4, pp. 3-25.

Livesey, Graham (2005): "Assemblage." In: Adrian Parr (ed.), The Deleuze Dictionary Revised Edition, Edinburgh: Edinburgh University Press, pp. 18-19.

Los Angeles Times (1999): "Owens Corning Fails to Block Some Blacks From Asbestos Suit", In: *Los Angeles Times* March 26, (http://articles.latimes.com/print/1999/mar/26/news/mn-21243) accessed May 17, 2012.

Lowry, Deborah W. (2004): "Understanding Reproductive Technologies as a Surveillant Assemblage: Revisions of Power and Technoscience" Sociological Perspectives 47/4, pp. 357-370.

Luhmann, Niklas (1997): Die Gesellschaft der Gesellschaft, Frankfurt am Main: Suhrkamp.

Lyotard, Jean-François (1984): The Postmodern Condition: A Report on Knowledge, Minneapolis: University of Minnesota Press.

MacAttram, Matilda (2009): "The DNA database betrays the racism of those behind it." In: The Guardian July 13, (http://www.guardian co.uk/society/joepublic/2009/jul/13/dna-database-black-community/) accessed January 15, 2012.

Mackenzie, Adrian (2005): "Problematising the Technological: The Object as Event?" Social Epistemology 19/4, pp. 381-399.

MacKenzie, Donald (1990): Inventing Accuracy, Cambridge, MA: MIT Press.

MacKenzie, Donald (1998): Knowing Machines. Essays on Technological Change, Cambridge, MA and London: MIT Press.

MacKenzie, Donald/Wajcman, Judy (1985): "Introductory Essay." In: Donald MacKenzie/Judy Wajcman (eds.), The Social Shaping of Technology, Milton Keynes: Open University Press, pp. 2-25.

Maddox, Brenda (2003): Rosalind Franklin: The Dark Lady of DNA, London: HarperCollins.
Marcuse, Herbert (2004 [1941]): "Some Social Implications of Modern Technology." In: Herbert Marcuse, Technology, War and Fascism: Collected Papers of Herbert Marcuse, Vol. 1, ed. by Douglas Kellner, London and New York: Routledge, pp. 41-65.
Marx, Karl (1920 [1847]): The Poverty of Philosophy, Chicago: Charles H. Kerr & Company.
Marx, Karl (1971): The Grundrisse, New York, San Francisco, and London: Harper & Row.
Marx, Karl (1976 [1867]): The Capital. A Critique of Political Economy, Vol. 1, London and New York: Penguin Books.
Marx, Karl (1993 [1858]): "The Fragment on Machine." In: The Grundrisse, London and New York: Penguin Books, pp. 690-712.
Marx, Karl (1998 [1932]): The German Ideology, Amherst, MI and New York: Prometheus Books.
Mason, Chris (2012): "Theresa May to split up UK Border Agency." In: *BBC News* February 20, (http://www.bbc.com/news/uk-politics-17099143/) accessed September 12, 2013.
Massumi, Brian (2004): "Translator's Foreword: Pleasures of Philosophy." In: Gilles Deleuze and Félix Guattari: A Thousand Plateaus, New York and London: Continuum.
Mbembe, Achille (2003): "Necropolitics." Public Culture 15/1, pp. 11-40.
Merchant, Carolyn (1980): The Death of Nature: Women, Ecology, and the Scientific Revolution, London: Wildwood House.
Miller, Kelly (1897): A Review of Hoffman's Race Traits and Tendencies of the American Negro, Washington D.C.: The American Negro Academy.
Misa, Thomas J. (1992): "Controversy and Closure in Technological Change: Constructing 'Steel'." In: Wiebe E. Bijker/John Law (eds.), Shaping Technology/Building Society. Studies in Sociotechnical Change, Cambridge, MA: MIT Press, pp. 109-139.
Mitropoulos, Angela (2016): "On Borders/Race/Fascism/Labour/Precarity/Feminism/etc." In: base October 29, (http://www.basepublication.org/?p=107/) accessed November 30, 2016.
Mol, Annemarie (2002): The Body Multiple: Ontology in Medical Practice, Durham and London: Duke University Press.
Mumford, Lewis (1967): The Myth of the Machine: Technics and Human Development, San Diego: Hartcourt, Brace & World.

Müller, Ruth/Kenny, Martha (2014): "Nature/Nurture Refigured: Gendered Narratives of Parenthood in Environmental Epigenetics." Paper presented at the annual meeting of the 4S Annual Meeting, Copenhagen Business School, October 20, 2014.

Murphy, Michelle (2012): Seizing the Means of Reproduction: Entanglements of Feminism, Health, and Technoscience, Durham and London: Duke University Press.

Nail, Thomas (2016): Theory of the Border, Oxford: Oxford University Press.

Nature (2009): "Editorial: Genetics without borders." Nature 461, pp. 697.

Nietzsche, Friedrich (1989 [1887]): On the Genealogy of Morals & Ecce Homo, ed. by Walter Kaufmann, New York: Vintage Books.

Oates, John (2009): "Home Office Declines to Detail DNA-for-Foreigns Trial." In: The Register September 30, (http://www.theregister.co.uk/2009/09/30/dna_aslyum/) accessed January 21, 2012.

Oudshoorn, Nelly/Pinch, Trevor (eds.) (2003): How Users Matter: The Co-Construction of Users and Technology, Cambridge, MA: MIT Press.

Owens Corning (2015): "Owens Corning Earns Top Marks in 2016 Corporate Equality Index", November 18 (http://media.owenscorning.com/press-release/corporate/owens-corning-earns-top-marks-2016-corporate-equality-index) accessed June 20, 2017.

Petty, Thomas (2002): "John Hutchinson's Mysterious Machine Revisited." Chest 121/5, pp. 219-223.

Pias, Claus (2004): Cybernetics: The Macy-Conferences 1946-1953. Volume 2: Documents, Zürich: diaphanes.

Pinch, Trevor (2011): "Karen Barad, Quantum Mechanics, and the Paradox of Mutual Exclusivity." Social Studies of Science 41/3, pp. 431-441.

Pinch, Trevor/Bijker, Wiebe E. (1984): "The Social Construction of Facts and Artefacts: or How the Sociology of Science and the Sociology of Technology Might Benefit Each Other." Social Studies of Science 14/3, pp. 399-441.

Pinch, Trevor/Ashmore, Malcolm/Mulkay, Michael (1992): "Technology, Testing, Text: Clinical Budgeting in the U.K. National Health Service." In: Wiebe E Bijker/John Law (eds.), Shaping Technology/Building Society: Studies in Sociotechnical Change, Cambridge, MA: MIT Press, pp. 265-289.

Plotnitsky, Arkady (2013): Niels Bohr and Complementarity: An Introduction, New York and London: Springer.

Reardon, Jenny (2012): "The Democratic, Anti-Racist Genome? Technoscience at the Limits of Liberalism," Science as Culture 21/1, pp. 25-47.

Rouse, Joseph (2004): "Barad's Feminist Naturalism" Hypatia 19/1, pp. 142-161.

Ruppel, Gregg L. (2006): "Which Reference Set for Spirometry?" FOCUS: Journal for Respiratory Care & Sleep Medicine, January, (https://www.highbeam.com/doc/1G1-143342126.html) accessed October 21, 2017.

Saldanha, Arun (2006): "Re-Ontologising Race: The Machinic Geography of Phenotype." Environment and Planning D: Society and Space 24/1, pp. 9-24.

Saldanha, Arun/Adams, Jason Michael (eds.) (2013): Deleuze and Race, Edinburgh: Edinburgh University Press.

Sarasin, Philipp (2006): Anthrax: Bioterror as Fact and Fantasy, Cambridge, MA: Harvard University Press.

Sawicki, Jana (1999): "Disciplining Mothers: Feminism and the New Reproductive Technologies." In: Janet Price/Margrit Shildrick (eds.), Feminist Theory and the Body, New York: Routledge, pp. 190-202.

Schaffer, Simon (1991): "The Eighteenth Brumaire of Bruno Latour." Studies in History and Philosophy of Science 22/1, pp. 174-192.

Scheper-Hughes, Nancy/Wacquant, Loïc (2002): Commodifying Bodies, London: Sage.

Schmidgen, Henning (2012): "Inside the Black Box: Simondon's Politics of Technology." SubStances 41/3, pp. 16-31.

Schrödinger, Erwin (2006 [1944]): What is Life?, Cambridge, MA: Cambridge University Press.

Serres, Michel (1982): The Parasite, Baltimore and London: The Johns Hopkins University Press.

Shachtman, Noah (2007): "Robo-Snipers, 'Auto Kill Zones' to Protect Israeli Borders." In: Wired June 2, (http://www.wired.com/2007/06/for_years_and_y/) accessed December 20, 2012.

Shapin, Steven/Schaffer, Simon (1985): Leviathan and the Air-Pump: Hobbes Boyle, and the Experimental Life, Princeton: Princeton University Press.

Simondon, Gilbert (1958): Du Mode d'existence des objets techniques, Paris: Aubier.

Simondon, Gilbert (2016): On the Mode of Existence of Technical Objects, Minneapolis: University of Minnesota Press.

Singer, Mona (2012): "Retro-Figuren des kulturell Anderen: Wider die kulturalistische Viktimisierung von Migrant_innen." In: Gender Initiativkolleg (ed.), Gewalt und Handlungsmacht: Queer_Feministsiche Perspektiven, Frankfurt am Main: Campus, pp. 181-195.

Singer, Mona (2015): "Und was sagt Eva? Warum die Feministin keine Transhumanistin sein will, Posthumanistin dagegen schon." Wespennest 169, pp. 50-54.

Sismondo, Sergio (1993): "Response to Knorr Cetina." Social Studies of Science 23/3, pp. 565-569.

Sismondo, Sergio (2004): An Introduction to Science and Technology Studies, Malden and Oxford: Blackwell.

Skoog, Douglas A./Holler, James F./Crouch, Stanley R. (2007). Principles of Instrumental Analysis, Belmont, CA: Thomson Brooks/Cole.

Sokal, Alan (1996): "A Physicist Experiments with Cultural Studies." Lingua France May/June, (http://linguafranca.mirror.theinfo.org/9605/sokal.html) accessed January 20, 2013.

Sokal, Alan (2008): Beyond the Hoax: Science, Philosophy and Culture, Oxford and New York: Oxford University Press.

Stiegler, Bernard (2018): The Neganthropocene, London: Open Humanities Press.

Stone, Allucquère Rosanne (Sandy) (2002): "Will the Real Body Please Stand Up? Boundary Stories about Virtual Culture." In: David Bell/Barbara M. Kennedy (eds.), The Cybercultures Reader, London and New York: Routledge, pp. 504-528.

Sullivan, Nikki (2012): "The Somatechnics of Perception and the Matter of the Non/Human: A Critical Response to the New Materialism." European Journal of Women's Studies 19 /3, pp. 299-313.

The Guardian (2013): "Cameron: I'd Withdraw from Human Rights Convention 'to Keep UK Safe'." In: *The Guardian* September 29, (https://www.theguardian.com/politics/2013/sep/29/david-cameron-human-rights-convention/) accessed October 27, 2013.

Thornhill, Randy/Palmer, Craig T. (2000): A Natural History of Rape: Biological Bases of Sexual Coercion, Cambridge, MA: MIT Press.

Travis, Alan (2009): "Police Told to Ignore Human Rights Ruling Over DNA Database." In: The Guardian August 7, (http://www.guardian.co.uk/politics/2009/aug/07/dna-database-police-advice/) accessed October 20, 2011.

Travis, John (2009): "Key Questions on Nationality Testing." In: Science Online September 29, (http://news.sciencemag.org/2009/09/key-questions-nationality-testing/) accessed February 24, 2012.

UK Border Agency (2009a): "Stakeholders Letter", September 11, (http://news.sciencemag.org/scienceinsider/entryassets/stakeholder%23letter.11.9 0 9.doc) accessed December 12, 2011.

UK Border Agency (2009b): Nationality Swapping – Isotope Analysis and DNA Testing, (http://news.sciencemag.org/scienceinsider/entry-assets/nationality-swapping-DNA-testing.pdf) accessed December 12, 2011.

UK Home Office (2011): "FOI release 20818 Human Provenance Pilot Project." Gov.UK, December 12, (https://www.gov.uk/government/publications/20818-human-provenance-pilot-project/) accessed March 5, 2012.

van der Ploeg, Irma/Sprenkels, Isolde (2011): "Migration and the Machine-Readable Body: Identification and Biometrics." In: Huub Dijstelbloem/Albert Meijer (eds.), Migration and the New Technological Borders of Europe, Basingstoke: Palgrave Macmillan, pp. 68-104.

van der Tuin, Iris (2011): "A Different Starting Point, a Different Metaphysics: Reading Bergson and Barad Diffractively." Hypatia 26/1, pp. 22-42.

van der Tuin, Iris (2014): "Diffraction as a Methodology for Feminist Onto-Epistemology: On Encountering Chantal Chawaf and Posthuman Interpellation." Parallax, 20/3, pp. 231-244.

van der Tuin, Iris/Hoel, Aud Sissel (2012): "The Ontological Force of Technicity: Reading Cassirer and Simondon Diffractively." Philosophy and Technology 26/2, pp. 187-202.

Vandenberghe, Frédéric (2002): "Reconstructing Humans: A Humanist Critique of Actant-Network Theory." Theory, Culture & Society 19/5-6, pp. 51-67.

Vannini, Phillip/Waskul, Denis (eds.) (2006): Body/Embodiment. Symbolic Interaction and the Sociology of the Body, Aldershot and Burlington: Ashgate.

Veblen, Thorstein (1914): The Instinct of Workmanship, and the State of Industrial Arts, New York: The Macmillan Company.

Wade, Nicholas (2014): A Troublesome Inheritance: Genes, Race and Human History, New York: Penguin Press.

Wajcman, Judy (1991): Feminism Confronts Technology, Cambridge: Polity Press.

Wajcman, Judy (2010): "Feminist Theories of Technology." Cambridge Journal of Economics 34/1, pp. 143-152.

Watson, James (1968): The Double Helix: A Personal Account of the Discovery of the Structure of DNA, New York: Atheneum.

Weber, Max (1978 [1922]): Economy and Society, Berkeley and Los Angeles: University of California Press.

Wehling, Peter (2006): "The Situated Materiality of Scientific Practices: Postconstructivism – a New Theoretical Perspective in Science Studies?" Science, Technology & Innovation Studies 1, pp. 81-100.

Weiss, Martin G. (2011): "Strange DNA: The rise of DNA Analysis for Family Reunification and its Ethical Implications." Genomics, Society and Politics 7/1, pp. 1-19.

Whittle, Andrea/Spicer, André (2008): "Is Actor Network Theory Critique?" Organization Studies 29/4, pp. 611-629.

Wiener, Norbert (1947): "A Scientist Rebels." Bulletin of the Atomic Scientists 3/1, p. 31.

Wiener, Norbert (1948): Cybernetics: Or Control and Communication in the Animal and the Machine, Cambridge, MA: MIT Press.

Wiener, Norbert (2003 [1954]): "Men, Machines, and the World." In: Noah Wardrip-Fruin/Nick Montfort (eds.), The New Media Reader, Cambridge, MA: MIT Press, pp. 67-72.

Wiener, Norbert/Pach, Leo (1983 [1946]): "From the Archives " Science, Technology, & Human Values 8/3, pp. 36-38.

Williams, Robin/Edge, David (1996): "The Social Shaping of Technology." Research Policy 25, pp. 865-899.

Wilson, Edward O. (1978): On Human Nature, Cambridge, MA: Harvard University Press.

Wilson, Elizabeth (1998): Neural Geographies: Feminism and the Microstructure of Cognition, New York: Routledge.

Winner, Langdon (1977): Autonomous Technology. Technics-out-of-Control as a Theme in Political Thought, Cambridge, MA: MIT Press.

Winner, Langdon (1986): The Whale and the Reactor: A Search for Limits in an Age of High Technology, Chicago and London: University of Chicago Press.

Winner, Langdon (1993): "Upon Opening the Black Box and Finding it Empty: Social Constructivism and the Philosophy of Technology." Science, Technology, & Human Values 18/3, pp. 362-378.

Winner, Langdon (2002): "Having the Technology. An Interview with Langdon Winner." Forth Door Review 6, pp. 104-109.

Winner, Langdon (2005): "Resistance is Futile: The Posthuman Condition and Its Advocates." In: Harold W. Baillie/Timothy K. Casey (eds.), Is Human Nature Obsolete? Genetics, Bioengineering, and the Future of the Human Condition, Cambridge, MA and London: MIT Press, pp. 385-411.

Wittgenstein, Ludwig (1953): Philosophical Investigations, Malden and Oxford: Blackwell.

Wolff, Megan (2006): "The Myth Of The Actuary: 'Life Insurance And Frederick L. Hoffman's Race Traits And Tendencies Of The American Negro'." Public Health Reports 121/1, pp. 84-91.

Woolgar, Steve/Cooper, Geoff (1999): "Do Artefacts Have Ambivalence? Moses' Bridges and Other Urban Legends in S&TS." Social Studies of Science 29/3, pp. 433-449.

Woolgar, Steve/Lezaun, Javier (2013): "The Wrong Bin Bag: A Turn to Ontology in Science and Technology Studies?" Social Studies of Science 43/3, pp. 321-340.

Wright, Thomas (2013): William Harvey: A Life in Circulation, Oxford and New York: Oxford University Press.

Social Sciences

Carlo Bordoni
Interregnum
Beyond Liquid Modernity

2016, 136 p., pb.
19,99 € (DE), 978-3-8376-3515-7
E-Book
PDF: 17,99 € (DE), ISBN 978-3-8394-3515-1
EPUB: 17,99 € (DE), SBN 978-3-7328-3515-7

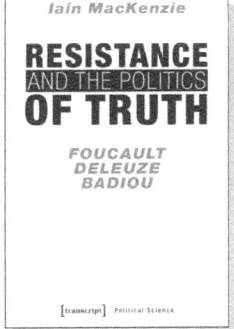

Iain MacKenzie
Resistance and the Politics of Truth
Foucault, Deleuze, Badiou

March 2018, 148 p., pb.
29,99 € (DE), 978-3-8376-3907-0
E-Book
PDF: 26,99 € (DE), ISBN 978-3-8394-3907-4
EPUB: 26,99 € (DE), ISBN 978-3-7328-3907-0

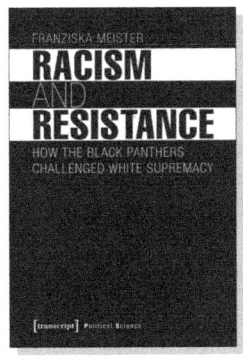

Franziska Meister
Racism and Resistance
How the Black Panthers Challenged White Supremacy

2017, 242 p., pb.
19,99 € (DE), 978-3-8376-3857-8
E-Book: 17,99 € (DE), ISBN 978-3-8394-3857-2

**All print, e-book and open access versions of the titles in our list
are available in our online shop www.transcript-verlag.de/en!**

Social Sciences

Cameron Harrington, Clifford Shearing
Security in the Anthropocene
Reflections on Safety and Care

2017, 196 p., hardcover
79,99 € (DE), 978-3-8376-3337-5
E-Book: 79,99 € (DE), ISBN 978-3-8394-3337-9

Benjamin Heim Shepard, Mark J. Noonan
Brooklyn Tides
The Fall and Rise of a Global Borough

February 2018, 284 p., pb.
29,99 € (DE), 978-3-8376-3867-7
E-Book: 26,99 € (DE), ISBN 978-3-8394-3867-1

Ilker Ataç, Gerda Heck, Sabine Hess, Zeynep Kasli, Philipp Ratfisch, Cavidan Soykan, Bediz Yilmaz (eds.)
movements. Journal for Critical Migration and Border Regime Studies
Vol. 3, Issue 2/2017: Turkey's Changing Migration Regime and its Global and Regional Dynamics

2017, 230 p., pb.
24,99 € (DE), 978-3-8376-3719-9

All print, e-book and open access versions of the titles in our list are available in our online shop www.transcript-verlag.de/en!